Socio-technical Systems in the Shaping and Development of Digital Business Models

Marek Jabłoński

CRC Press
Taylor & Francis Group
Boca Raton London New York

CRC Press is an imprint of the
Taylor & Francis Group, an **informa** business
A SCIENCE PUBLISHERS BOOK

First edition published 2025
by CRC Press
2385 NW Executive Center Drive, Suite 320, Boca Raton FL 33431

and by CRC Press
4 Park Square, Milton Park, Abingdon, Oxon, OX14 4RN

CRC Press is an imprint of Taylor & Francis Group, LLC

Library of Congress Cataloging-in-Publication Data (applied for)

ISBN: 978-1-032-59821-5 (hbk)
ISBN: 978-1-032-59822-2 (pbk)
ISBN: 978-1-003-45642-1 (ebk)

DOI: 10.1201/9781003456421

Typeset in Palatino
by Prime Publishing Services

Dedication

I dedicate this book to Professor Henryk Arctowski and Professor Antoni Bolesław Dobrowolski Polish polar explorers and members of the famous Belgica scientific and research expedition, considered the world's first climate scientists

Preface

The aim of the monograph is to demonstrate different approaches to building business models, where the starting point is the joint examination of social and technical factors. The aim is to demonstrate how today's technology allows us to build digital platforms based on communities. In this way, business ecosystems are created without state borders, which, thanks to technological mergers, improves their performance in terms of effectiveness and efficiency.

The monograph also aims to indicate the core aspects of designing and shaping digital business models based on the assumptions of the concept of socio-technical systems. The book presents a multidimensional approach to shaping digital platforms in the context of the determinants of complexity theory, business ecosystems and the taxonomy of various approaches to designing digital transformation solutions. Attention was paid to how modern technology allows for the creation of digital platforms based on communities. This creates the ecosystems of business ecosystems that, through technological mergers, improve the regulatory and business performance of digital business models. Socio-technical models and engineering techniques are used to search for solutions that are optimal in terms of the expected features of business solutions. Technology and models thereof stimulate the emergence of innovative solutions in the business space. Socio-technical and economic models represent a hybrid approach to designing configuration-based digital business models. The traditional business model was an organizational solution, but now organizational issues are being displaced by socio-technical models based on systems engineering and business digitalization. Cybernetics-based understanding, where the digital business model is based on the assumptions of system control refined by improvements resulting from the use of information through the use of multiple feedback loops, is experiencing a comeback. The management of such models is supported by service-oriented technologies. The concept of product and service systems is a crucial approach to the triad based on the use of socio-technical models, a feedback loop with users and the configuration of the digital business model. The issue posed in this way creates opportunities for the emergence of new formulas for shaping and developing technical solutions conducive to the creation of new functionalities that ensure the expected performance of digital business models based on scalability and value capture from the market. The book encompasses expertise in systems theory, cybernetics and a modern approach to shaping digital business models based on product–service systems. Innovative algorithms and socio-technical and engineering models support the design of digital business models. Product–service systems (PSS) models and socio-technical models are conducive to creating effective business solutions based on a high level of scalability. In this way, an interdisciplinary approach to shaping digital business models, which facilitates the development of a new, coherent idea of combining technology and management into a single whole, emerges. Management according to this approach is not a separate discipline, but instead depends on the technical solutions that are adopted for a given business solution. This idea is presented in this monograph. The book is a completely new approach based on a hybrid model combining technical and economic models in the design and development of digital business models based on socio-technical systems. A new perspective from the perspective of the achievements of cybernetics theory in the context of creating innovative solutions in the sphere of the digital economy, taking into account the role of social and ecological factors, triggers a new approach to creating digital platforms and explaining the details and reasons for their design. It is important to emphasize the design layer in the context of achieving the expected results, performance and functions of digital business models.

In order to illustrate the complexity of the issue at hand, chapters of the monograph and components of the author's original concept of identification of the conceptualization and operationalization of the mechanisms of socio-technical systems in the shaping and development of digital business models are presented graphically in Figure 1.

The book is preceded by this introduction. The first chapter broadly describes socio-technical systems as a research subject in management sciences. It presents core research assumptions and the cognitive gap in the context of solving a scientific problem and explains the basic assumptions of socio-technical systems in relation to current scientific achievements in this field. Further in this chapter, assumptions for the various forms of functioning of digital platforms—taxonomy—are developed. For the purpose of conducting research into the relationship between the concept of socio-technical systems and the design of digital

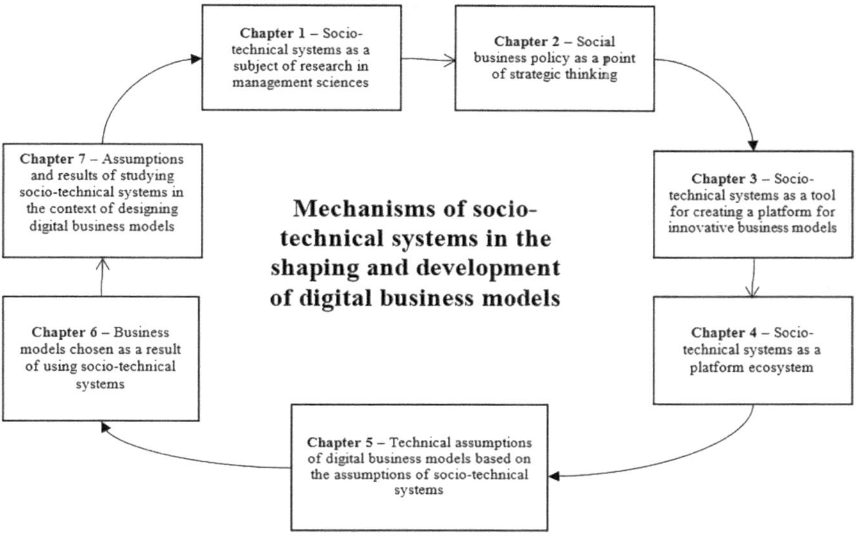

Figure 1 Presentation of the structure of the monograph: Mechanisms of socio-technical systems in the shaping and development of digital business models. *Source:* Own study.

business models, including business models in the form of digital platforms, a study was conducted by means of systematic literature review. Scientific articles and grey literature were used. A systematic review of the literature was conducted, which resulted in obtaining final conclusions and a report in accordance with the PRISMA standard. The second chapter focuses on socio-business policy issues in the context of conditions for shaping business ecosystems, in particular in the context of a multidimensional approach to the concept of scalability of digital business models. It indicates substantive areas such as socio-business policy—the place and role thereof in shaping business ecosystems, scalability and green scalability for social enterprises, the green scalability of digitally assisted circular business models, the scalability of technical, economic and social systems, the organizational scaling of the companies for social impact and a proposal of patterns with their characteristics describing the green scalability of digital business models based on the assumptions of the circular economy concept. The third chapter indicates the conditions for the construction of socio-technical systems in modern business. This scope of work was operationalized by clarifying cognitive issues such as social business ecosystems, the social aspects of business models, social community business models with a positive impact on the economy, the network paradigm and socio-technical systems and social enterprises in digital business ecosystems.

This chapter refers closely to the issue of identifying the place and role of social and ecological factors in the creation of attractive and supportive ideas operationalized in the form of digital platforms on the business ecosystems market in the digital economy. The fourth chapter is devoted to explaining issues related to socio-technical systems as a platform ecosystem, including areas such as a socio-technical perspective on the Digital Era, a business ecosystem in the digital reality, a technical ecosystem as a platform to build a network of connections, artificial intelligence and digital business—the sociotechnical approach and safety and cybersecurity as a factor in socio-technical system stability. This chapter aims to clarify definitions and different approaches to digital business ecosystems, which is particularly important to an understanding of how electronic markets work and under what conditions competitive advantages are built therein. The chapter contains content referring to a theoretical–evaluating study, including issues related to the technical assumptions of digital business models based on the assumptions of socio-technical systems. It comprises issues such as socio-technical models, algorithms and engineering methods as tools for shaping digital solutions in the digital economy. New business cybernetics concepts are understood as a tool for identifying and shaping core components of digital business models. Their role in controlling technical systems is focused on achieving the expected functionalities and performance which is important for servicing mass customers, using the servitization concept for digital transformation. The dynamic process of social and economic changes of business models is based on innovative solutions, automation and robotics. The processes based on shaping virtual digital business models and digital platforms based on algorithms and the Big Data environment in light of technical models based on innovation and traditional approaches are compared. The sixth chapter contains a description of selected business models as a result of using the socio-technical system. In addition, the issues of complex systems and the modularity of systems—complex products and complex organization, knowledge codification and fusion strategy for complex modularity systems for business platforms—were identified. Characteristic models include regenerative business models, sustainable digital business models, circular business models, progressive business models and inclusive business models. These characteristics are important for further research and analytical work in the context of assumptions for the design of digital platforms. The last chapter of the monograph, chapter seven, is of a research–conceptual nature, where several important models were developed in the course of research and scientific discussion. Various theoretical and application solutions related to the issue of digital platforms examined through the prism of the assumptions of socio-technical systems are presented. The chapter is entitled "The assumptions and results of research into socio-technical systems in the context of designing digital business models". The research results are preceded by a detailed description of the theoretical assumptions of research into socio-technical systems in the context of digital business models. The results of research and analysis include qualitative studies and case studies, including socio-technical systems—case studies for designing digital platforms as a business model, socio-technical systems as a platform for designing a railway system—the network approach, socio-technical systems as a platform for designing water and sewage sector companies and socio-technical systems as a platform for designing energy companies in the energy sector. These four case studies detail the place and roles of socio-technical systems in four digital business model creation environments which are important to an understanding of complex business ecosystems as part of digital transformation processes. The author's original canvas of the description of the digital business model based on the assumptions of socio-technical systems, as well as the original methodology for designing digital business models—digital platforms based on the assumptions of the concept of socio-technical systems—were developed. In addition, a set of management and technical–operational factors that have the greatest impact on the efficiency of the digital business models was defined. The chapter concludes with a discussion of the research results—conclusions for theory, practice, future directions of research and limitations of the research process, theoretical implications, practical implications, potentially interesting future directions of research and limitations of the research, and strategic recommendations for the use of socio-technical systems in the shaping and development of digital business models. The substantive content of the monograph ends with a conclusion that synthetically summarizes the content of the monograph. The last part of the monograph consists of references, a list of tables and figures.

Marek Jabłoński

Contents

List of Tables

List of Figures

Chapter 1

Socio-technical Systems as a Subject of Research in Management Sciences

1.1 INTRODUCTION

Socio-technical systems are examined in science from many different perspectives. They are the subject of research not only from the perspective of social sciences, but also technical and engineering sciences. As regards management sciences, this concept is one of the core paradigms underlying other solutions, especially in the area of systems theory. Management, as an applied science oriented towards the application of the proposed solutions, draws on the experience of the authors of numerous scientific papers dealing with this concept. It is particularly important to consider the concept of socio-technical systems from the perspective of designing business models, in particular digital business models. Social and technical systems also shape business ecosystems because, especially in the area of the digital economy, the assumption is to build communities which combine the same interests, passions and needs based on innovative modern technologies. The design and management of digital business models should be supported by solutions based on the combined implementation of social and technological aspects. Thus, theoretical core assumptions underlying the operation of digital platforms and the creation of their functions and performance depend mainly on variables that relate to the systems theory and assumptions of combining social and technological factors.

DOI: 10.1201/9781003456421-1

1.2 SOCIO-TECHNICAL SYSTEMS IN THE RELEVANT LITERATURE

Socio-technical theory deals with the interdependent design of the social system and the technical system so that they operate for the benefit of the organization (Cooper and Foster 1971). The social subsystem includes people, the relationships between them, formal principles, values, goals and structures (Koźmiński and Piotrowski 2006). Socio-technical systems have a rich history of development in this area of knowledge. The assumptions of socio-technical systems concern the integration of social and technical elements. The core principle is that these systems combine technology with human social elements. Designers focus on designing systems based on how people interact with technology. The evolution of this concept dates back to the 1950s. The founder of this concept is Eric Trist, who, as a psychologist working on several projects related to the mining sector in Great Britain, formulated the idea of a socio-technical system and described the company as a social and technical workplace (Trist and Bamforth 1951). He referred these analyses to workplace activities in search of synergies between technology and social factors. Socio-technical systems can be understood as the systematic integration of two types of phenomena that have very divergent or even partially contradictory characteristics. Socio-technical systems consist of computers, networks and software, as well as people, procedures, policies, laws and many other aspects. They need to be co-designed so that they can be used effectively by synergy (Fischer and Herrmann 2011). According to Ropohl, engineers in particular tend to ignore social concerns about their work, while social scientists, on the other hand, do not know a great deal about technology and are reluctant to take into account the artificial reality of technical objects (Ropohl 1999). The complexity of this concept proves that various approaches to this subject, characterized by concepts such as socio-technical theory, the socio-technical system and socio-technical system theory, emerge in the relevant literature. Socio-technical systems are said to be efficient and useful when they are 'optimized together', but few system designers understand what joint optimization is and how to achieve it (Di Maio 2014). Another issue is skillful risk management as part of achieving the usability of these systems. Managers responsible for complex socio-technical systems will need a risk management framework for complex systems—a framework that can provide decision support within operationally relevant timeframes (Luther et al. 2013). Socio-technical systems (STS) should be designed as agile work systems that respond to environmental challenges while institutionalizing the core organizational functions that enable value creation and retention in an organization (Botla and Kondur 2018). Socio-technical systems theory adopts an understanding of the enterprise system in which the balance, mutual optimization and compatibility of both technical and social systems are considered with equal attention, in addition to careful consideration of the situation at work and the well-being of employees, as well as the productivity of work and efficiency (Fayoumi and Williams 2021). Socio-technical systems belong to the group of complex systems characterized by features such as system size/number of elements, the variety of elements, connections, disorder, dynamics, information content, uncertainty and system structure (Schöttl and Lindemann 2015). In recent years, in the context of the dynamics of conducting the processes of digital transformation of economic systems, this is important in the context of defining assumptions for socio-technical systems. For the transformation to be successful, social and technical systems need to be jointly optimized (Govers and van Amelsvoort 2019). Then, the meta design of socio-technical systems should be consistent with the need to integrate two types of structures and processes, namely technical systems that are designed to ensure predictable and reliable interactions between users and systems and social systems that are dependent in their interactions and are subject to evolution (Fischer and Herrmann 2011). It should be noted that modern socio-technical systems must be based on Big Data sets. Big Data Analytics (BDA) is an emerging technology for handling Big Data. The use of advanced techniques with varying levels of intelligence, such as descriptive, predictive, prescriptive and autonomous analysis, is expected to create new value for companies (Duong Oesterreich et al. 2022). The use of socio-technical systems is very broad. One of the other applications of such systems is to use them to identify and solve problems related to crisis management. Many models can be used to mitigate crisis events, for example, numerous very interesting adaptations of B. Turner's model of the development of prediction to prevent future events (Turner 1976). This solution was analyzed, among others, in order to search for solutions to neutralize the effects of the COVID-19 pandemic (Borodzicz 2020). In another context, an understanding of the contemporary role of socio-technical systems can draw attention to the landscape in the field of social media and the role of these systems in their functional support. They offer new possibilities and even condition their existence (Lombardo et al. 2021). Another application is the use of this concept to design solutions based on Lean Management. Although lean production is usually associated with a reduction of complexity, it is increasingly used in highly complex socio-technical systems. It is interesting to hypothesize the relationship between a simplification strategy using system complexity theory (Soliman and Saurin 2017). Socio-technical systems theory, described as complex systems, are also designed taking into account the use of the Internet of Things, e.g., in decentralized energy supply systems in smart cities. Complex systems are then deployed and operated by means of centralized computing infrastructures. Then, the socio-technical nature of these decentralized systems requires new design paradigms sensitive to values that strengthen trust, transparency and compliance with social values (Fanitabasi et al. 2020). Socio-technical systems are also used to design systems that are safe in terms of safety engineering (Sheikh Bahaei et al. 2021). They also allow for the search for new risk assessment formulas in terms of their attributes which are conducive to the modeling of safety and risk management processes (Teixeira de Souza et al. 2022). In addition, an interesting trend emerges in the approach to risk management in safety-related industries, which focuses on the effective use of adaptations in complex socio-technical systems (Fostera et al. 2019). The use of the Rasmussen model of socio-technical systems (Rasmussen 1997) is common in identifying potential mitigation/prevention measures that could improve risk assessment from higher levels in the system (Hunt and Naweed 2023). Another interesting approach to the innovative application of the assumptions of socio-technical systems is their use in projects which deal with the problem of climate change and adaptation strategies, but also counteracting these processes. Designing sectoral transformations of socio-technical systems requires going beyond an economic and technological focus towards a broader view of the system, combining social, behavioral and institutional elements with natural environments and infrastructures (Hoicka et al. 2022). Even problems in proving fault in the use of complex systems, whether they concern people or technical systems, are addressed in the relevant literature (Brahim et al. 2021). Practical solutions are also sought for the implications of the processes of effective design and implementation of IT systems using the role of socio-technical systems. This approach may be based on a focus on optimizing the role of human factors in the design of these systems (Lawler et al. 2011). Moreover, a qualitative and quantitative approach to resilience engineering for the management of socio-technical system safety is crucial, taking into account the assessment of parameters such as effectiveness, efficiency, flexibility and resilience (Eljaoed et al. 2020). Also, in the process of designing and analyzing the application of artificial intelligence, the important role of socio-technical systems in ensuring a comprehensive and intelligent approach must not be forgotten. It is then necessary to design the entire system of work taking into account the approach of the socio-technical system (Gabriel et al. 2022). Socio-technical systems are also increasingly examined in the context of servitization issues. Digital servitization means moving from delivering products to offering intelligent systems comprised of products and services. The adoption of a socio-technical systems perspective ensures the interrelationship and complementarity of internal and external elements relevant to the delivery of an intelligent product–service system (PSS) (Münch et al. 2022).

Attention should be paid to the important relationship in the understanding of the use of socio-technical systems. Both the 'technical' and 'social' worlds are systemic when they interact. Together they also become new systems on their own terms. For B. Turner, this interaction between social and technical systems was the place where most of the failures occurred (Turner 1976). The relationship between the social system and the technical system in the context of the identification of space for the creation of digital business models in the business ecosystem is presented in Figure 1.1.

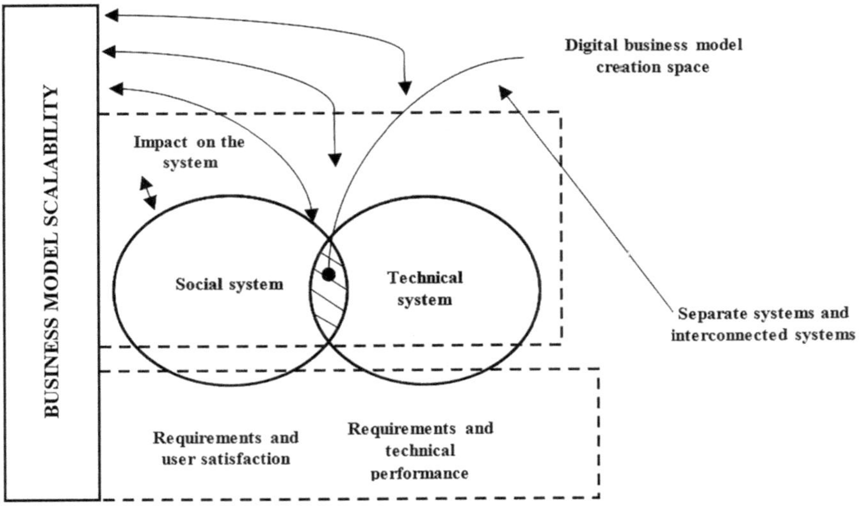

Figure 1.1 The relationship between social systems and the technical system in the context of designing an ecosystem of digital business models.
Source: Own study.

The presented diagram is based on the assumption that the space for creating digital business models based on digital platforms is made by creating a social system and a technical system and combining them in order to obtain an appropriate business environment. It can be called a business ecosystem based on the assumptions of the integration of the social system with the technical system. Only the two systems combined create the conditions for creating new value for users and monetizing the business. The common space is the most efficient in this respect and is a necessary condition for designing digital business models. The five levels of the taxonomy of the factors that describe socio-technical systems are shown in Figure 1.2.

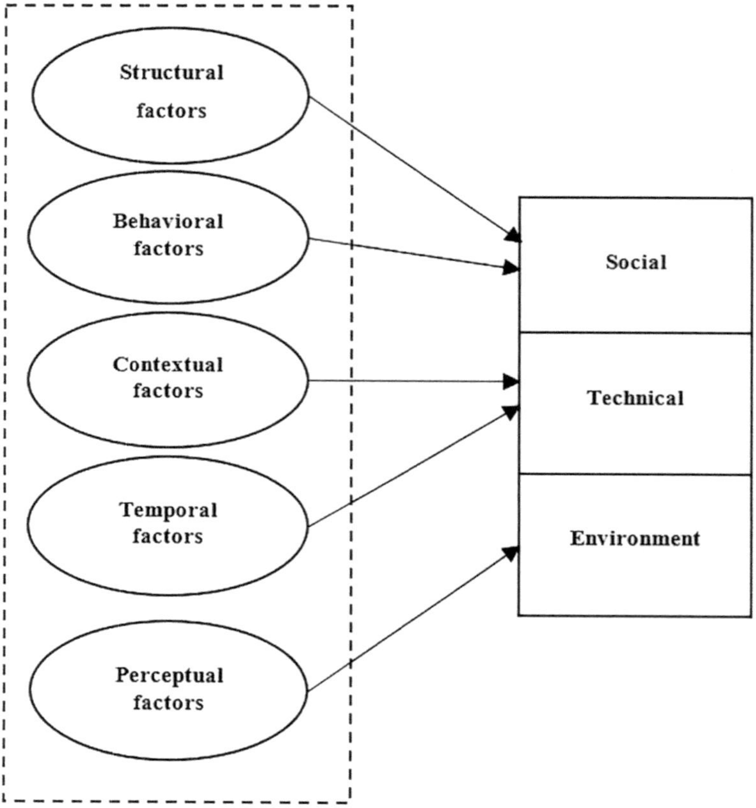

Figure 1.2 Five levels of socio-technical system taxonomy.
Source: Own study based on (Rhodes and Ross 2010) Shaping Socio-Technical System Innovation Strategies Using a Five Aspects Taxonomy. Available online: http://seari.mit.edu/documents/preprints/RHODES_ EUSEC10.pdf (accessed on 26.12.2024).

In this model, every aspect of the taxonomy is equivalent to every dimension of the socio-technical system, namely the social, technological and environmental areas. All factors, namely structural factors, behavioral factors, contextual factors, temporal factors and perceptual factors, are included in these three leading areas of the design of these systems. Each of these thematic areas should be integrated into a unified system based on the principles of complexity theory.

1.3 FORMS OF FUNCTIONING OF DIGITAL PLATFORMS–TAXONOMY

Today's digital platforms are very diverse. They are based on numerous variables that result from market and technological parameters, resulting from the range of impact, income generation logic, local conditions, interfaces between mobile devices and technological fusions. The logic of digital platforms depends on technological and business issues that at the same time make up a specific business model. The relationship between these two variables is crucial and management processes cannot be examined only from the perspective of traditional management. Technology determines the conditions for doing business, enables interaction between network actors and ensures that a specific innovative solution can be implemented. Only when technological issues are clarified and developed in terms of a certain level of reliability and information security can a layer of business logic be imposed that takes into account customer segmentation, their needs and market structure. Therefore, it is reasonable to state that the effectiveness and efficiency of digital platforms depends on understanding the problem of complexity of systems and imposing the principles of functioning of socio-technical systems on it. Therefore, technical issues, interfaces between devices, interactions between users, the possibility of value delivery, value capture and integrating the platform into the economic system are conditioned by technological business models such as digital platforms. Digital technologies make business projects more efficient due to the multifunctional approach to shaping them. The results of digital technology are more accurate than traditional solutions (Andadari et al. 2021). It is possible to show contemporary trends multidimensionally through taxonomy, which is a tool of scientific cognition. Therefore, there is a need to develop a taxonomy of digital platforms due to their specificity in the context of digital business models, which is described in the following subsection.

Taxonomy is a method of classifying and describing entities characterized by systematics. It is a systematic division of a given group of issues. In this case, this concerns the concept of digital platforms. An attempt to systematize concepts in this area is necessary to fully understand the principles of functioning of digital business models based on platforms. Platforms, their functions and archetypes were divided according to numerous different criteria. Table 1.1 presents a multidimensional taxonomic analysis of digital platforms based on a critical literature analysis.

Table 1.1 Taxonomy of Digital Platforms

No.	Type of Digital Platform	Description
		Definitions (Blaschke et al. 2019)
	Platform Owner	Natural or legal entity that designs, implements, and maintains the digital platform
	Third Party	Natural or legal entity that augments the technical core with complementary derivatives (e.g., software extensions, services, and sales channels)
	End User	Natural or legal entity that uses the resources available on the digital platform
	Digital Ecosystem	Complex network of platform-mediated actor-to-actor interactions, which becomes increasingly accessible to end users through third parties' platform derivatives
	Service	Specialized competences (knowledge and skills) exchanged among different actors in the digital ecosystem through deeds, processes, and performances
	Technical Core	Extensible codebase serving as a building block upon which third parties devise platform-augmenting derivatives
	Digital Infrastructure	The computing and network resources that allow distributed actors to facilitate their resource exchange across spatial, temporal, and organizational boundaries
		Dimensions and characteristics of digital platforms (Freichel et al. 2021)
	Architecture	Digital architecture integrating various operating systems on the basis of which specific software for creating digital platforms is designed.
	Scope	The scale and scope of digital platforms are clearly defined. The scope can be defined from the perspective of the geographical coverage as well as the target audience segment.
	Business Model	The way the enterprise creates and delivers value to customers, and then converts payments received to profits (Teece 2018).
	Value Proposition (Osterwalder and Pigneur 2010) Revenue Model Cost Model	Product and Service; Customer Needs; Geography Pricing Logic; Channels; Customer Interaction Core Assets and Capabilities; Core Activities; Partner Network
	Control	Control includes control of the value chain, understood as a way to attract generative platform-related activities.

No.	Type of Digital Platform	Description
	Functionality	Functionality includes the way a system performs specific functionalities—it can be shared by modules and interfaces.
	Price Strategy	The price strategy includes the way in which the value of the product or service offered to customers is determined—it depends on a valuation based on profit margin.
	Revenue Source	Revenue sources include all methods of generating revenue by providing a certain amount of value to buyers of products or services.
	Openness/Access	Open access includes the possibility of full, unrestricted access for network users of a given piece of information, data, digital product, etc.
	Involvement	The strategic involvement of key stakeholders supporting a given digital platform.
	Trust	Trust is a variable accepted by recipients and can be understood as the belief that a given platform will perform the expected actions of value to the user, and at the same time the risk of using this platform and the potential loss of this value is acceptable.
	Market Structure	It covers relationships between suppliers and recipients of value and determines the structure of sales that can be carried out in a specific area.
	C2C	A consumer-to-consumer relationship that describes the type of business model where transactions are made directly between consumers.
	B2C	The B2C model refers to the relationship between a company and individual consumers. A specific company sells products or services directly to end users.
	B2B	A B2B (business-to-business) agreement is a form of civil law agreement concluded between two different companies. Both parties to the contract are equal.
	Division of digital platforms by T. Derave, F. Gailly, T. Prince Sales, G. Poels (Derave et al. 2024)	
1.	One-sided platform	A one-sided digital platform encompasses only one group of recipients from whom it derives value. An example of a single digital platform is the B2C eCommerce business model.
2.	Two-sided platform	A two-way digital platform allows interaction between two different user groups. The value creation effect depends on the participation of both groups at the same time. The platform connects buyers and sellers. Two-sided platform models help

Contd.

No.	Type of Digital Platform	Description
		two parties to get in touch and solve their problems. Typical examples of two-sided markets are Uber or Airbnb.
3.	Multi-sided platform	A multi-sided platform brings together interested people around a specific and shared goal. It does this to bring people together to create value in the marketplace where the community is created. This community is focused on common interests and even shared ideas.
4.	Transaction platform	Trading platforms are technical, digital solutions that support the exchange of goods, services or information between multiple networked recipients—users or participants. They connect buyers and sellers based on the network effect. Platforms operate on the principle of scalability. The more users are involved in the platform, the more value is generated for it.
5.	Investment platform	An investment platform is an electronic trading platform which, through a digital network, is used to place orders for financial products. It is a digital platform-based intermediary in the execution of financial transactions.
6.	Crowdfunding platform	Crowdfunding platforms are used for charitable, investment, corporate finance and personal needs. Many people use crowdfunding to donate to numerous charitable causes. They are technological and digital platforms that are widely available to donors who want to support a project.
7.	Digital marketplace	The digital marketplace is an online marketplace or electronic sales platform. It is a type of website that allows sellers and buyers to carry out commercial transactions. This is an online marketplace where different sellers can offer their products or services. Potential customers can view, compare and buy them.
8.	Peer-to-peer (P2P) sharing and collaborative consumption platform	Peer-to-peer is a network that is mainly used to share files between Internet users by means of the appropriate application. The digital resources made available include music files, movies, computer programs or multimedia presentations. Such a system is built from a decentralized network architecture in which participants, called peers, communicate directly with each other without the need for a central authority or server.

No.	Type of Digital Platform	Description
9.	Sharing economy platform	The sharing or collaborative economy means business models in which activity takes place through the mediation of collaborative platforms. They create a generally available market for the temporary use of goods or services. They are provided by private individuals. The sharing economy comprises three categories of participants: • service providers sharing their goods, resources, time or skills. These may be natural persons offering services on an occasional basis or service providers professionally engaged in the provision of services; • service users; • intermediaries who, through an online platform, connect providers with users.
10.	On-demand platform	A value delivery business model in which computational resources are made available to the user on demand. This type of service is provided using methods such as virtualization, computer clusters and distributed processing.
11.	Second-hand P2P platform	An online platform for the direct trading of second-hand goods used by private individuals. It is a digital platform for the exchange of second-hand goods.
	From linear markets to platforms and ecosystems (How platforms create value for their users: implications for the Digital Markets Act, Prepared for the Computer and Communications Industry Association 2021, www.oxera.com.)	
	Linear Markets	Traditional markets are characterized by a linear supply chain, where manufacturers sell to retailers—wholesalers who then sell those goods to consumers ("retail").
	Three-sided platforms	A digital three-sided platform is a business model that is based on enabling transactions between external producers, consumers and intermediaries. This is how value is created.
	Dual-mode operators	A business model that aims to provide a layer of protection and stability to computer systems by separating user programs and the operating system into two modes: user mode and kernel mode. User mode restricts access to privileged resources, while kernel mode has full access to those resources.
	Ecosystems	A digital environment where commercial transactions are carried out, but also interactions between the actors of the Internet network. It is built through many subsystems based on a specific architecture.

Contd.

No.	Type of Digital Platform	Description
		Four archetypes of digital platforms (Staub et al. 2021)
	Business Innovation Platform	A platform that implements innovative, non-standard functions based on business digitalization. It breaks current rules to create new value through innovative methods.
	Consumer Innovation Platform	Specialized innovation management software that offers companies innovative ways to collect, refine, implement and monitor new ideas in order to generate a rapid return on investment (ROI) through added value to customers. It covers tasks such as co-creation, user-feedback, iterative design, customization and personalization, and technology integration.
	Business Exchange Platform	A platform for the exchange of value between producers and customers. Technology serves as a bridge between them.
	Consumer Exchange Platform	A technological solution based on a digital platform for the exchange of value between customers through the supply of producers and their mutual exchange of value.
		Archetypes of digital platform transactions (Analysis and methodology developed by Marissa Dean, Caribou Digital, 2018 (CC BY-NC-SA 4.0), https://www.financedigitalafrica.org/wp-content/uploads/2019/06/Digital-Platforms-Business-Model-Analysis-Guide.pdf [accessed on 30.12.2024]).
	E-commerce	Trade-based activity of buying and selling goods and services over the Internet. Customers from the e-commerce sector can make purchases using mobile devices.
	Global Online Work	A type of remote work that can be performed anywhere in the world if the operator has access to a computer and an Internet network.
	Local Services	Local services within a small geographical area based on local culture and consumer behavior.
	Asset Sharing	An asset-sharing business model that combines spare capacity and demand, allowing consumers to access a product or asset rather than forcing them to own it.
	Attention	A business model where the value seller is a person who provides services in the crowd attraction industry, creating something hard to ignore, and then selling that attention to advertisers.
	Date	These are sets of values that convey information, describing quantity, quality, fact, statistics, other meanings or sequences of symbols that can be further interpreted and processed.

No.	Type of Digital Platform	Description
	Configurations of digital platforms for manufacturing (den Hartigh et al. 2023)	
	Internal platform	An internal platform for managing internal production processes and all functions and process performance along with the flow of necessary documents.
	Supply chain platform	A platform for managing value chain activities, including demand planning, production planning, sourcing and supplier management, supply network planning, detailed scheduling, logistics, and supply chain analytics.
	Industry Platform	A platform for managing processes as part of the exchange of knowledge and information between market participants, including cooperation between production and service facilities.
	Key concepts in the digital platform context (Blaschke et al. 2019)	
	Platform Owner	Natural or legal entity that designs, implements, and maintains the digital platform
	Third Party	Natural or legal entity that augments the technical core with complementary derivatives (e.g., software extensions, services, and sales channels)
	End User	Natural or legal entity that uses the resources available on the digital platform
	Digital Ecosystem	Complex network of platform-mediated actor-to-actor interactions, becoming increasingly accessible to end users through third parties' platform derivatives
	Service	Specialized competences (knowledge and skills) exchanged among different actors in the digital ecosystem through deeds, processes, and performances
	Technical Core	Extensible codebase serving as a building block upon which third parties devise platform-augmenting derivatives
	Digital Infrastructure	The computing and network resources that allow distributed actors to facilitate their resource exchange across spatial, temporal, and organizational boundaries
	Platform types (Freichel et al. 2021)	
	Payment	These are platforms for making electronic payments.
	Electronic Market	An electronic market that ensures that economic activities are carried out by means of digital economy solutions based on multidimensional links between different suppliers and recipients of value via the Internet.

Contd.

No.	Type of Digital Platform	Description
	App Store	The App Store is a digital app distribution platform, created and managed by Apple Inc., designed for the iOS, iPadOS, watchOS and tvOS operating systems.
	Archetype of digital platform (Blaschke et al. 2019)	
	Orchestration Platform	Digital platforms that bring together federated networks—open, vertically distributed and open ecosystems—of third parties expanding the platform through cooperative and inclusive platform profiles.
	Amalgamation Platform	Digital platforms that create private networks—self-centered, vertically integrated and closed ecosystems—involving an exclusive selection of several private actors through monopoly and assimilation platform profiles.
	Innovation Platform	These are digital platforms that provide unrestricted access to new digital infrastructure without rights.
	Basic platform types—examples (Cusumano et al. 2020)	
	Transaction platforms	• Snapchat • Instagram • Twitter • Airbnb • JD.com • Uber • Lending Club • TripAdvisor • Amazon Marketplace • WeChat • Facebook • Salesforce Exchange • Steam • Windows Store • Alibaba • Rakuten • LinkedIn • Match.com • Baidu • Google Play • Pinterest • Apple App Store
	Hybrid companies	• Apple • Google • Microsoft • Valve • Salesforce • Facebook • Tencent • Amazon

No.	Type of Digital Platform	Description
	Innovation platforms	• Apple iOS • ARM CPU • IBM Watson • Nintendo • Intel CPU • Sony PlayStation • GE Predix • SAP NetWeaver • Qualcomm Brew • Google Android • Microsoft Azure • Steam Machine • Force.com • Facebook for Developers • WeChat APIs • Amazon Web Services

Source: Own study based on the relevant literature quoted in the table.

1.4 SYSTEMATIC LITERATURE REVIEW—THE DESCRIPTION OF THE SOCIO-TECHNICAL SYSTEMS RESEARCH LANDSCAPE IN THE CONTEXT OF DESIGNING DIGITAL BUSINESS MODELS/DIGITAL PLATFORMS

For the purpose of conducting research into the relationship between the concept of socio-technical systems and the design of digital business models, including business models in the form of digital platforms, research was conducted by means of a systematic literature review. Scientific articles and grey literature were used. The systematic review of the literature resulted in final conclusions and a report produced in accordance with the PRISMA (Preferred Reporting Items for Systematic Reviews and Meta-Analyses) standard (Tricco et al. 2018). The research gap is the identification of the degree to which socio-technical systems theory is combined with the concept of digital business models. This topic has not yet been addressed in the relevant literature, although it seems surprising considering the logical potential of the assumptions of the concept of socio-technical systems. The research procedure comprised the following steps:

1. Gathering the relevant literature on socio-technical systems, digital business models and digital platforms.

2. Conducting a systematic literature review by means of the Web of Science and Scopus database.
3. Developing a report for systematic literature reviews and meta-analyses—application of the PRISMA standard.
4. Analyzing findings based on the results of research conducted as part of the systematic literature review.

In the relevant literature, the concept of a socio-technical system is characterized and described in depth. However, the aim of this systematic literature review was to examine whether and to what extent it occurs alongside the concepts of a digital business model and a digital platform. Do the scientific achievements accumulated in the scientific literature support the processes of studying and designing digital business models, including digital platforms? Based on two leading databases of scientific publications, i.e., Web of Science and Scopus, a search for the term 'socio-technical system' in the Business, Management and Accounting category brought quantitative results in this respect in terms of the common occurrence of these concepts in the scientific literature. Studies showed a fairly small quantitative association of these concepts in the relevant literature. In the first place, the Scopus and WoS databases were searched for the terms **Sociotechnical Systems versus Digital Business Model** and **Sociotechnical Systems versus Digital Platform.** The decision was also made to search for the terms "Sociotechnical Systems" and "Digital Business Model" and "Digital Platform" to have a broader perspective, and here it turned out that 12 entries were obtained in Scopus and 11 entries in WoS. These are not large numbers. After removing the duplicates, 20 items were left. Based on these studies, the PRISMA model was used and a report on the results of the systematic literature review was prepared in this form. Figure 1.3 presents the PRISMA report for given concepts.

The results indicating selected scientific articles strongly linking defined concepts in order to assess their interdependence are presented in the tables below.

The Table 1.2 shows the list of publications after the Scopus and WoS database search for the term **"Sociotechnical Systems versus Digital Business Model".**

The Table 1.3 shows the list of publications after searching the Scopus and WoS databases for the term **"Sociotechnical Systems versus Digital Platform".**

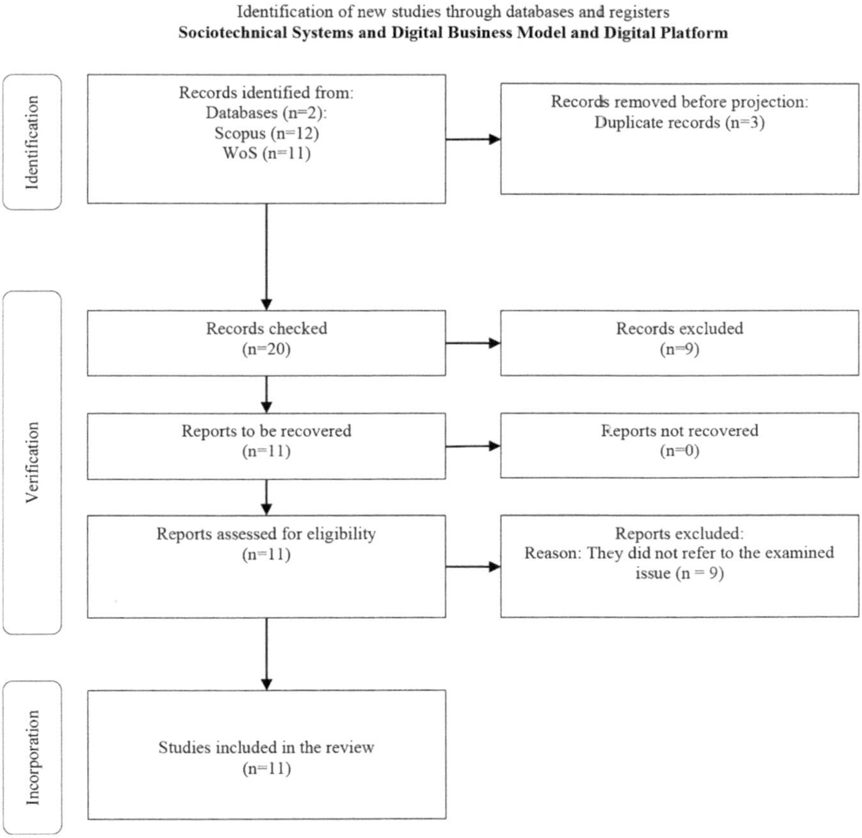

Figure 1.3 Report on the methodology of the systematic literature review
Source: Own study.

Table 1.2 List of Publications after Scopus and WoS Database Search for the Term "Sociotechnical Systems versus Digital Business Model"

	Sociotechnical Systems versus Digital Business Model	Title	Authors	Year	Document Type
Scopus	1	Digital innovation and the future of work	H. Schaffers, M. Vartiainen, J. Bus	2020	Book
WoS	1	Transparency as design choice of open data contests	S. Brunswicker, B. Jensen, Z.N. Song, A. Majchrzak	2018	Article

Source: Own study.

Table 1.3 List of Publications after Scopus and WoS Database Search for the Term "Sociotechnical Systems versus Digital Business Model"

	Sociotechnical Systems versus Digital Platform	Title	Authors	Year	Document Type
Scopus	2	An Exploration of Dutch Dermatologists' Experience and Satisfaction With Teledermatology: Sociotechnical and Complex Adaptive System Perspective	F. van Sinderen, C. Kuziemsky, M.W. Jaspers, L.W., Peute	2024	Article
		Digital innovation and the future of work	H. Schaffers, , M. Vartiainen, J. Bus	2020	Book
WoS	0	–	–	–	–

Source: Own study.

In the subsequent step, the keywords included in the 11 specialist articles included in the study were presented, and a semantic map of the publications on "Sociotechnical Systems versus Digital Business Model" and "Sociotechnical Systems versus Digital Platform" in the area of "business, management and accounting" was created for those keywords included in the content of those articles.

Based on the research, a detailed identification of keywords related to the research (Table 1.4) was made possible.

Table 1.4 Identification of Keywords Related to the Research

Id	Keyword	Occurrences	Total Link Strength
1	agile manufacturing systems	1	20
2	agile transformational approach	1	20
3	artificial intelligence	1	9
4	autonomous agents	1	14
5	bridges	1	12
6	Business and Societies	1	20
7	business development	1	20
8	business models	3	54
9	business opportunities	1	20
10	business transaction	1	12
11	circular economy	1	16
12	circular ecosystem	1	16
13	circular ecosystems	1	16
14	classification (of information)	1	12
15	cognition	1	20

Id	Keyword	Occurrences	Total Link Strength
16	cognitive mapping	1	20
17	cognitive systems	1	20
18	competitive analysis	1	18
19	complex data	1	16
20	complexity theory	1	14
21	computational experiment	1	14
22	computational methods	1	14
23	computational model	1	14
24	consumer preferences	1	14
25	content creation	1	18
26	content creators	1	18
27	crowd business models	1	20
28	crowdfunding	1	18
29	customer satisfaction	1	20
30	decision support systems	1	20
31	design	1	16
32	design science research	1	16
33	design-science researches	1	16
34	digital business ecosystems	1	14
35	digital innovation	1	9
36	digital platform ecosystem	1	16
37	digital platform ecosystems	1	16
38	digital platforms	2	36
39	digital technologies	1	20
40	digital transformation	1	20
41	digitalization	1	9
42	domain model	1	12
43	economics	1	12
44	ecosystems	2	30
45	electronic data interchange	1	16
46	embedded systems	1	18
47	employment	1	9
48	enhanced socio-technical (est) model	1	20
49	ethics	1	9
50	facebook	1	18
51	finance	1	18
52	financial support	1	18
53	functional features	1	12
54	future of work	1	9
55	future prospect	1	20
56	game-changing technologies	1	9
57	groupware	1	18

Contd.

Id	Keyword	Occurrences	Total Link Strength
58	industrial development	1	20
59	information systems	1	14
60	information use	1	14
61	innovation	2	34
62	interactive computer systems	1	18
63	interpretive structural modeling	1	20
64	interpretive structural modeling (ism)	1	20
65	interpretive structural models	1	20
66	large dataset	1	12
67	mapping	1	20
68	metadata	1	20
69	new business opportunity	1	20
70	online users	1	12
71	open innovation	1	9
72	operational environments	1	12
73	patreon	1	18
74	patronage	1	18
75	platform design	1	14
76	platforms of ecosystems	1	16
77	platform work	1	9
78	requirements engineering	1	12
79	sales	1	20
80	simulation and modeling	1	14
81	mac	1	20
82	smacit	1	20
83	smart city	1	20
84	social media	1	18
85	social networking (online)	1	18
86	social support	1	18
87	society 5.0	1	20
88	socio-economics	1	12
89	socio-technical approach	1	20
90	sociotechnical	3	50
91	sociotechnical system design	1	16
92	sociotechnical systems	3	48
93	sociotechnical systems design	1	16
94	software design	1	20
95	strategic planning	1	20
96	structural design	1	14
97	systemic effect	1	20
98	systems analysis	1	20
99	systems engineering	1	20

Id	Keyword	Occurrences	Total Link Strength
100	technological development	1	20
101	technological evolution	1	20
102	technological innovation	1	20
103	transformational approach	1	20
104	user goals	1	12
105	utility type digital platforms	1	20
106	workplace innovation	1	9
107	youtube	1	18

Source: Own study

Semantic map of publications on the concepts of "Sociotechnical Systems versus Digital Business Model" and "Sociotechnical Systems versus Digital Platform" in the area of "business, management and accounting" are shown in Figure 1.4.

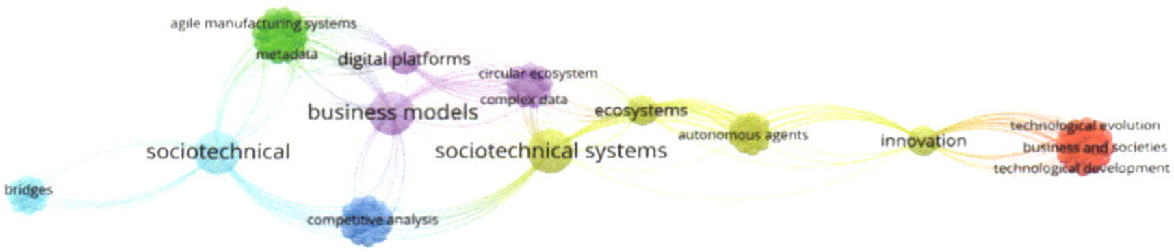

Figure 1.4 Semantic map of publications on the concepts of "Sociotechnical Systems versus Digital Business Model" and "Sociotechnical Systems versus Digital Platform" in the area of "business, management and accounting". *Source:* Own study by means of the VOSviewer tool.

The generally evaluated scientific literature identified by keywords using the Scopus and WoS databases is new and refers to the latest trends in the theory and practice of management and business. Unfortunately, it is clear that researchers do not make much use of the significant number of achievements in the area of social and technical systems, while digital platforms bringing together communities and connecting them through technological solutions clearly meet the definition of social and technical systems. From a logical point of view, there should be no doubt in this regard. Platforms build a business ecosystem in which communities exchange mutual values, and the only tool that enables them to do so is technological digital platforms. The following concepts appear around the concepts of the socio-technical system and business model: digital platforms, ecosystems, innovation, competitive analysis, circular ecosystems, complex data, autonomous agents, technologies and evolutions, business and societies, technologies development, agile manufacturing systems and metadata. These are very new concepts in the field of scientific and practical cognition. This therefore proves that the concept of socio-technical systems is increasingly being examined in the context of new trends in the digital economy and the design of digital business models. A total of 107 words from the reviewed articles also indicate that this topic is very modern, innovative and progressive. Numerous concepts require a great deal of research and analyses on their impact on the principles and cycles of designing digital business models and digital platforms. In addition to the basic concepts, the issues of innovation and the business ecosystem, which are raised later in the monograph, emerge. The presented concepts indicate that this topic is developing dynamically and these processes should be studied, as the assumptions of socio-technical systems and scientific achievements accumulated in the field of this topic are becoming a leading concept that supports the understanding and design of digital solutions based on technologies and the active participation of the community on the basis of feedback loops.

1.5 CONCLUSIONS

The theoretical assumptions of the functioning of technical social systems in the context of determinants of management processes presented in the chapter have proven that this topic is part of the issue of designing and developing digital business models, in particular digital platforms, where socio-technological issues should be examined together. They determine the investment attractiveness of business models, business efficiency and effectiveness, as well as the ability to achieve the assumed business goals. The analysis and bibliographic research based on the methodology of the systematic literature review have shown that despite the intuitive understanding of socio-technical systems and in-depth literature research, modern literature draws little from the experience resulting from research in the field of systems theory, including socio-technical systems. The impact of socio-technical systems on the design of digital business models is fundamental to achieving the expected results and a better understanding of this issue from the perspective of management theory and practice.

REFERENCES

Analysis and methodology developed by Marissa Dean, Caribou Digital. 2018. (CC BY-NC-SA 4.0). https://www.financedigitalafrica.org/wp-content/uploads/2019/06/Digital-Platforms-Business-Model-Analysis-Guide.pdf [dostęp 30.12.2024].

Andadari, T.S., Purwanto, L.M.F., Satwiko, P. and Sanjaya, R. 2021.. Study of digital architecture technology: theory and development. Journal of Architectural Research and Education. 3(1): 14–21.

Blaschke, M., Haki, K., Aier, S. and Winter, R. 2019. Taxonomy of Digital Platforms: A Platform Architecture Perspective, 14th International Conference on Wirtschaftsinformatik, Siegen, Germany.

Borodzicz, E.P. 2020. The Utility of a 'Socio-Technical' systems theory of Crisis? *In*: Proceedings of the 6th International Workshop on Socio-Technical Perspective in IS Development.

Botla, L. and Kondur, H. 2018. Socio Technical systems of a company: The dimensionality of socio technical systems. PURSHARTHA. XI(1).

Brahim, A., Kyriakopoulos, S. and Pretschner, A. 2021. Causality-based accountability mechanisms for socio-technical systems. Journal of Responsible Technology. 7–8.

Cooper, R. and Foster, M. 1971. Sociotechnical systems. American Psychologist. 26(5): 467–474. doi:10.1037/h0031539.

Cusumano, M.A., Yoffie, D.B. and Gawer, A. 2020. The future of platforms, platforms power the world's most valuable companies, but it will get harder and harder to capture and monetize their disruptive potential. MIT Sloan Management Review. 61(3). Reprint #61304. https://mitsmr.com/2Szn7Tj.

den Hartigh, E., Stolwijk, C.C.M., Ortt, J.R. and Punter, L.M. 2023. Configurations of digital platforms for manufacturing: an analysis of seven cases according to platform functions and types. Electronic Markets. 33(1), Article 30. https://doi.org/10.1007/s12525-023-00653-4.

Derave, T., Gailly, F., Prince Sales, T. and Poels, G. 2024. A taxonomy and ontology for digital platforms. Information Systems. 120: 6. https://doi.org/10.1016/j.is.2023.102293.

Di Maio, P. 2014. Towards a Metamodel to Support the Joint Optimization of Socio Technical Systems. 273–296. doi:10.3390/systems2030273.

Duong Oesterreich, T., Anton, E., Teuteberg, F. and Dwivedi, Y.K. 2022. The role of the social and technical factors in creating business value from big data analytics: a meta-analysis. Journal of Business Research. 153: 128–149.

Eljaoed, W., Yahia, N.B. and Saoud, N.B.B. 2020. A qualitative-quantitative resilience assessment approach for socio-technical systems. Procedia Computer Science. 176: 2625–2634.

Fanitabasi, F., Gaere, E. and Pournaras, E. 2020. A self-integration testbed for decentralized socio-technical systems. Future Generation Computer Systems. 113: 541–555.

Fayoumi, A. and Williams, R. 2021. An integrated socio-technical enterprise modelling: A scenario of healthcare system analysis and design. Journal of Industrial Information Integration. 23: 100221.

Fischer, G. and Herrmann, T. 2011. Socio-technical systems: A meta-design perspective. International Journal of Sociotechnology and Knowledge Development. 3(1): 1–33.

Fostera, C.J., Plant, K.L. and Stanton, N.A. 2019. Adaptation as a source of safety in complex socio-technical systems: A literature review and model development. Safety Science. 118: 617–631.

Freichel, C., Fieger, J. and Winkelmann, A. 2021. Developing a taxonomy for digital platforms—A conceptual approach. *In*: Proceedings of the 54th Hawaii International Conference on System Sciences. 5782.

Gabriel, S., Bentler, D., Grote, E.M., Junker, C., Meyer zu Wendischhoff, D., Bansmann, M., et al. 2022. Requirements analysis for an intelligent workforce planning system: a socio-technical approach to design AI-based systems. Procedia CIRP. 109: 431–436.

Govers, M. and van Amelsvoort, P. 2019. A socio-technical perspective on the digital era: The lowlands view. European Journal of Workplace Innovation. 4(2). Special Issue.

Hoicka, Ch. E., Zhao, Y., McMaster, M.-L. and Das, R.R. 2022. Diffusion of demand-side low-carbon innovations and socio-technical energy system change. Renewable and Sustainable Energy Transition. 2: 100034.

Hunt, D. and Naweed, A. 2023. The risk of risk assessments: Investigating dangerous workshop biases through a socio-technical systems model. Safety Science. 157.

Koźmiński, K. and Piotrowski, W. (red). 2006. Zarządzanie: teoria i praktyka [Management: Theory and Practice], wyd. 5, Wydawnictwo Naukowe PWN, Warszawa. 30–31.

Lawler, E.K., Hedge, A. and Pavlovic-Veselinovic, S. 2011. Cognitive ergonomics, socio-technical systems, and the impact of healthcare information technologies. International Journal of Industrial Ergonomics. 41.

Lombardo, G., Mordonini, M. and Tomaiuolo, M. 2021. Adoption of social media in socio-technical systems: A survey. Information. 12: 132.

Luther, B., Gunawan, I. and Nguyen, N. 2023. Identifying effective risk management frameworks for complex socio-technical systems. Safety Science. 158: 105989.

Münch, Ch., Marx, E., Benz, L., Hartmann, E. and Matzner, M. 2022. Capabilities of digital servitization: Evidence from the socio-technical systems theory. Technological Forecasting and Social Change. 176.

Osterwalder, A. and Pigneur, Y. 2010. Business model generation: A handbook for visionaries. Game Changers, and Challengers. Wiley, Hoboken, NJ.

Rasmussen, J. 1997. Risk management in a dynamic society: A modelling problem. Safety Science. 27(2/3): 183–213.

Ropohl, G. 1999. Philosophy of socio-technical systems. PHIL & TECH. 4: 3. Spring.

Schöttl, F. and Lindemann. U. 2015. Quantifying the complexity of socio-technical systems—A generic. Interdisciplinary Approach, Procedia Computer Science. 44: 1–10.

Sheikh Bahaei, S., Gallina, B. and Vidović M. 2021. A case study for risk assessment in AR-equipped socio-technical systems. Journal of Systems Architecture. 119.

Soliman, M. and Saurin, T.A. 2017. Lean production in complex socio-technical systems: A systematic literature review. Journal of Manufacturing Systems. 45: 135–148.

Staub, N., Haki, K., Aier, S. and Winter, R. 2021. Taxonomy of digital platforms: A business model perspective. *In*: Proceedings of the 54th Hawaii International Conference on System Sciences. 6168.

Teece David, J. 2018. Business models and dynamic capabilities. Long Range Planning. 51: 40–49.

Teixeira de Souza, I., Rosa, A.C., Patriarca, R. and Haddad, A. 2022. Soft computing for nonlinear risk assessment of complex socio-technical systems. Expert Systems with Applications. 206.

Tricco, A.C., Lillie, E., Zarin, W., O'Brien, K.K., Colquhoun, H., Levac, D., et al. 2018. Annals of Internal Medicine. PRISMA Extension for Scoping Reviews (PRISMA-ScR): Checklist and Explanation. 169(7): 467–473.

Trist, E.L. and Bamforth, K.W. 1951. Some social and psychological consequences of the longwall method of coal getting. Human Relations. 4: 3–38.

Turner, B. 1976. The development of disasters—A sequence model for the analysis of the origins of disasters. Sociological Review. 24(4): 753–774.

Chapter 2

Social Business Policy as a Point of Strategic Thinking

2.1 INTRODUCTION

Social policy, which plays a crucial role in the process of designing digital business models, especially those oriented towards social impact, is a key point of strategic thinking about business. Ecological and economic aspects are fundamental concepts that designers of digital business models need to pay attention to. This chapter provides an in-depth analysis of the concept of scalability from technological, economic and ecological perspectives. Attempts were made to explain all these three concepts in terms of the determinants of modern business ecosystems focused on the aspects of digital technologies and other trends such as the greening of digital platforms. Attempts were also made to explain why the scalability of business models is so vital and why the success or failure of digital platforms depends on it, as well as what determinants should be used to unambiguously indicate the factors that affect this scalability in the process of designing, building and using digital business models.

2.2 SOCIAL BUSINESS POLICY—ITS PLACE AND ROLE IN SHAPING BUSINESS ECOSYSTEMS

Modern business ecosystems are shaped by technical social systems that define the role of technology and humans in creating value for recipients and other stakeholder groups. The socio-technical system

DOI: 10.1201/9781003456421-2

is created through synergy between hardware, software, people and communities. It is based on explaining the role of social structures, roles, and laws (social sciences) in the context of designing systems that target communities of people and the application of specific technologies. The concept of the ecosystem exists in at least two key thematic areas, namely strategic management and innovation and technology management (ESSEC Chair of Networked Business Models, Business Ecosystems: a Literature Review, Working Paper, Economics of Mutuality Alliance, 2017). A business ecosystem is a network system that connects parts of the real and digital worlds that interact and integrate with each other. A business ecosystem is also a system composed of entities that need each other and exist within various groups of interests as a community (Gao 2021). Moore's work on business ecosystems reveals two approaches to this phenomenon: one analogous, the other definitional. His work was based on the assumption of an analogy of business in ecology (Moore 1993, 1996, 2006). Therefore, the term 'ecosystem' is generally used in the field of biological sciences in the ecology section. However, according to the development of science and technology, the science of the ecosystem has been introduced into economic and business literature. Ecology and the economy share many similarities. Ecology in this sense can be considered a natural economy that carries out transactions in the form of materials, energy and information (Noviyanti and Tricahyono 2024). Various analogies arise for defining a business ecosystem, including those of a biological ecosystem, as well as an industrial ecosystem, an economy as an ecosystem, a digital business ecosystem and a social ecosystem (Peltoniemi and Vuori 2004). Research and analysis of business ecosystems cover at least three key areas, namely focusing on the dynamics of the ecosystem, especially in contacts with stakeholders rather than its structure. Second, researchers should conduct more research embedded in relationships rather than research into logic that prevails in the context of value-based management. Finally, future research should extend to the international context and not be limited to specific technologies at the local level (Rong et al. 2018). Digital ecosystems, on the other hand, operate as systems where a group of representatives works collectively on the basis of co-opetition, i.e., they cooperate on the basis of their own and shared rules and infrastructures, and independently compete to create and offer services to their clients, regardless of their structure and nature (Espina-Romero et al. 2023). This type of ecosystem is defined as an economic community, comprising many companies working together to gain a comparative advantage as a result of their symbiotic relationships. Ecosystems then allow companies to create new values that no company could achieve alone (Kim et al. 2010). A crucial role in this respect is played by social business policy, where communities and individuals of people are most important in terms of business design. Technological support serves people and the usefulness of the products and services they design. A social business policy is created by conscious business designers who pay special attention to the positive social impact. This impact is relevant in the sense that doing business is not only profit-oriented, but in particular helps people and communities to improve their quality of life. Such a modern approach is aimed at supporting modern technologies based on innovative algorithms often implemented in real time, taking into account artificial intelligence. They are created through technology fusions, where enterprises enter into a relationship of supporting their technologies by building networks of data, information, software, hardware and interfaces between them. An integral part of this technological sphere is communities of people, mainly users, who create a coherent synergy with technologies in the technological environment. Social business policy is therefore crucial to shaping business ecosystems. Network- and fusion-based technology requires a range of agreements between providers of these solutions in the service of business and social value creation. A business ecosystem that takes into account social sensitivity supported by technologies must be based on cooperation between people in large communities. The joint implementation of the assumptions of the social impact and the network effect creates the conditions for new business opportunities. In this view, the business ecosystem is based on the participation of large communities of people using IT applications designed on the principle of technology fusion of core players in digital business. This applies not only to large corporations such as Facebook, Google or Amazon, but also to other players who understand that, firstly, today's business is based on cooperation, and secondly, the space for new technologies will be its use by a business partner as part of its technological solution. Based on such a philosophy, the know-how of the product should be preserved, but its usability and performance should be used as a value for another supplier. Barriers to the development of such an approach are the fear some economic entities have of losing intellectual property, which results in their loss of competitiveness. By securing intellectual property while opening up to cooperation, it builds technology fusion opportunities that underlie the creation of business ecosystems in terms of digital platforms. Social policy, value economy and social impact allow for the creation of a new approach based on the success of all business players, creating added value for societies which often results in an improvement in quality of life while maintaining the ability to create profits; but these profits, which are important, are socially accepted by stakeholders, which is conducive to a long-term development strategy. A condition for the development of this solution is synergy between people, between technologies and between people and technologies. Understood in this way, the business ecosystem allows for the creation of new, revolutionary business models. Policy in this situation, focused on social and positive reception by communities of people and understood as a set of principles that managers follow when making managerial decisions, should shape the modern business ecosystem based on technological platforms. Figure 2.1 presents a multidimensional arrangement of links between the business model and the ecosystem based on the assumptions of the socio-technical system.

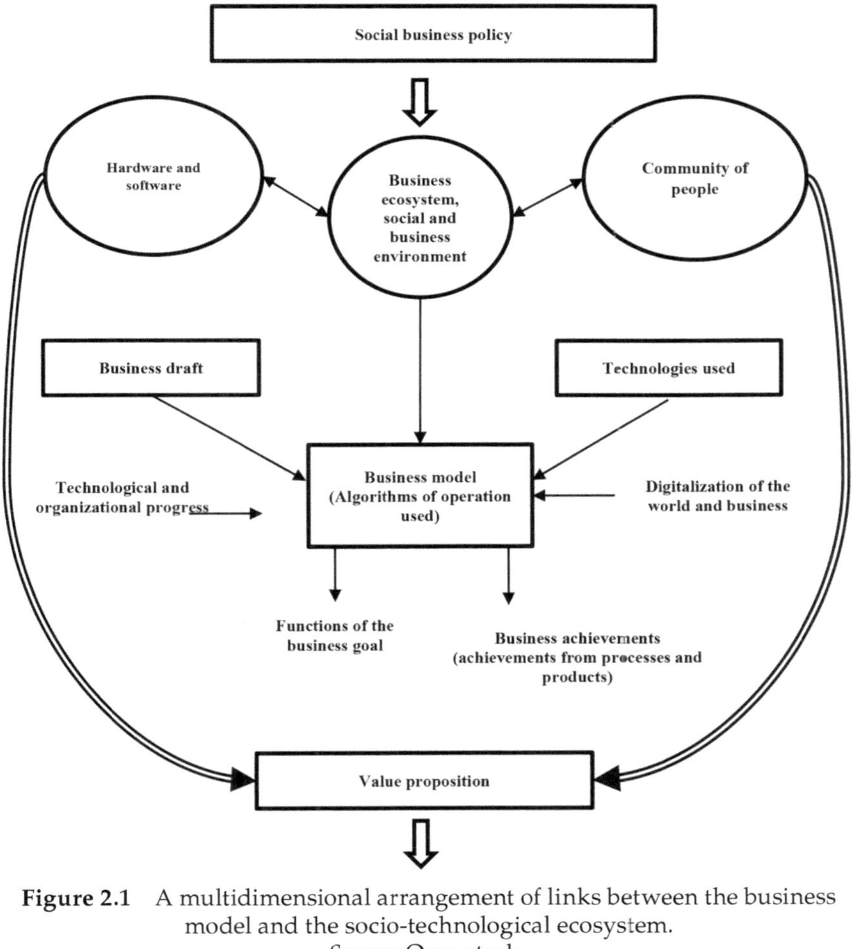

Figure 2.1 A multidimensional arrangement of links between the business model and the socio-technological ecosystem.
Source: Own study.

The multiplicity of relationships and variables indicates that it is a complex system. Many systems and subsystems interact with each other, strengthening the core system of the company's business model, the primary goal of which is to deliver value propositions to customers and users. Technological progress, business digitalization and the use of specific technologies and the business context are the determinants of the development of the business model. The social and technological environment shapes the possibilities of functioning of modern business models. Against this background, specific business goals and results deriving directly from the functions and performance of processes and products are achieved.

2.3 SCALABILITY AND GREEN SCALABILITY FOR SOCIAL ENTERPRISES

Social enterprises play a key role in the modern economy. A social enterprise is one that sets a positive social impact as its primary objective, taking into account the ability to operate effectively. It can be said that the economy needs social enterprises in order to fully satisfy the needs of citizens. The traditional approach of enterprises to profit maximization is beginning to be replaced by a more conscious approach that meets the high expectations of modern societies. Quality of life, health and safety at work, responsibility and citizen satisfaction can only be ensured if companies take into account the needs of stakeholders in their strategies, but in the sense that such activities will be part of daily operational activities and long-term strategic goals. The scalability of social enterprises is a very interesting concept, because the author believes that individual initiatives affect a very wide audience. Then, a relationship occurs that refers to the concept of a multiplier of the scale of operation of social enterprises. When a given initiative significantly affects the social environment and the social environment needs this initiative for its development, we can talk about the scalability of a social enterprise. When the social impact significantly exceeds the expenditure on the implementation of a specific social program, the scalability effect is achieved. A social enterprise is not scalable when a number of actions undertaken by the social enterprise do not specifically translate into specific social outcomes expected by stakeholders. Thinking based on scalability is based primarily on social initiatives that have a chance of broad public perception, satisfy the most important social needs, improve the quality of life of citizens and improve the efficiency of processes of civic services. This also applies to those areas of the economy where the state has a smaller share and people's needs are captured from the bottom up and implemented through specific social programs. Scalability understood in this way is crucial to the development of modern regions and countries.

2.4 GREEN SCALABILITY OF DIGITALLY ASSISTED CIRCULAR BUSINESS MODELS

The contemporary conditions of business in the conditions of the growing role of those stakeholders who expect business activity to have a positive impact on the natural environment create the need to assess the impact of pro-ecological activities from a point of view which is broader than the area of microeconomic prospects. This requires a holistic view far exceeding the assessment of the direct impact of an individual company on its close environment. In this respect, the problem lies in how to assess the impact of an individual company on climate change. A way to find the answer to this question may be to explain the concept of green scalability, the effectiveness of which can be assessed through the prism of the circular economy and circular business models built on its foundations. The starting point of this concept may be the assumption that the relationship between the resources involved in the production or provision of services and the ecological effects will be significantly positive. In other words, the added value will not only be the effect of a linear value chain, but of a circular process value loop. In this aspect, the single process of creating added value will be multiplied by further possibilities of reduction, reuse, recycling and other mechanisms of this concept. If the environmental impact of a single company is combined with that of other companies operating in the same and related sectors, these circular business models will work together to create value for the wider public, and thus, as it were, for society as a whole. Such interdependent behavior could consequently have a certain impact on climate change, including positive changes.

It is crucial to develop the conceptualization and operationalization of the concept of green scalability of digitally assisted circular business models.

It is important to establish the relationship between the factors that affect the process of scalability of business models through circular digital transformation. The characteristics of the theoretical and application model of the process of transforming business models leads to the transition from a linear approach to a value loop. In this way it further explains the concept of green scalability. The current state of knowledge in this area is not complete. The issue of the green scalability of business models of companies operating in the circular economy has not been addressed in the relevant literature. In addition, the number of articles which deal with this issue is negligible, and the way they are described is fragmentary. It is crucial to clarify whether the scalability of business models in the area of the circular economy contributes to a significant reduction in the costs of production and service processes to a greater extent than other technologies and solutions, which should be crucial to improving the effectiveness and efficiency of business operations. Since digital transformation encompasses all processes, systems and work organization, which consequently changes the structure and logic of the business model, it should be noted that there is a significant cognitive gap in assessing the scalability of business through technical, technological, economic and organizational solutions and mechanisms of the circular economy.

2.4.1 Ecological Stimulators of Change in the Business Models of Modern Companies

Dynamic climate change resulting in negative weather anomalies is a good basis for the concern of the earth's inhabitants and experts for the future of our planet. This concern is also pragmatic if, instead of merely paying lip service to it, some entrepreneurs search for new formulas for running a business based on the neutralization and even stimulation of the positive effects of halting climate change. The inclusion of ecological factors as enhancers of certain components of business models creates opportunities such that their way of creating and delivering value will contribute to the creation of value for climate change neutralization. A significant number of entrepreneurs, seeing noticeable climate change, but also social expectations of changes in this area, develop their projects based on these needs. These are needs not only based on standard value drivers, such as sales growth or profitability, but in particular those which take into account the future possibilities of the social and ecological impact. Such innovative business models embedded in this space have a very good chance of achieving above-average investment attractiveness, which should directly translate into a high market capitalization value (Jabłoński 2015b). When analyzing and determining the determinants responsible for this investment attractiveness, ecological factors are increasingly taken into account, and even constitute an important growth stimulus. They are understood as areas which actively influence the perception of the business model by stakeholders interested in the positive ecological impact and pro-ecologically oriented customers seeking the creation of value aimed at this area from their suppliers. Built in the configuration of the business model and based on the idea of positive climate change, they fall into the range of those business models that address modern climate problems.

2.4.2 Green Digital Business Models—Assumptions of Ecological Digitalization

Digital transformation as an idea is conducive to optimizing the impact of manufacturing processes on the natural environment. The technological nature of digital business models makes access to these solutions much better than in the case of traditional business models. Another factor is the large extent of the impact, which is an important element in achieving a high level of business scalability. There are two reasons why green digital business models can be called so. Firstly, digital technology itself is based on optimizing the consumption of natural resources. Secondly, these business models are focused on creating ideas that support the implementation of positive climate policy. Green digitalization therefore has attributes built in both technology and in popular ideas attractive to environmentally friendly consumers. These two areas make it possible to create such defined green digital business models. The social environment expects entrepreneurs responsible for the environment to undertake projects. They are currently the most desirable direction for the development of early-stage companies. The combined implementation of the assumptions of the digital economy and environmental responsibility stimulates the emergence of green digital business models.

2.5 SCALABILITY OF TECHNICAL, ECONOMIC AND SOCIAL SYSTEMS

Scalability expresses a certain ability of an object to perform specific tasks. It is therefore dynamic and depends on the scale of operation of a given company or technical system. In technical terms, scalability describes the level of ensuring increasingly efficient operation as the number of components of the technical system increases. Scalability in a technological context describes the ability of a system to cope with an increased workload (Hill 1990).

Scalability in economic terms is a key parameter which determines the company's ability to grow, based, among other things, on the claim that not every unit of revenue is generated by an equal unit of cost. Thus, it can be defined as the ability to increase revenues faster when the corresponding cost base increases with fewer dynamics compared to these revenues (Hallowell 2001). As a result, investors often look for "infinite scalability", in which case an investment in the low profitability phase can bring high profits to the company, becoming highly profitable after amortizing fixed costs by servicing a large customer base (Stampfl et al. 2013). This is often referred to as growth hacking (Cavallo et al. 2023).

When scalability is examined in the context of the concept of business models, it is the ability of the business model to maintain similar or higher efficiency while constantly increasing or decreasing the number of its components, and at the same time constantly adjusting the limits of its impact (e.g., in a network environment). Scaling in a business model therefore involves, among other things, adding or subtracting a component and/or components of the business model in order to improve its effectiveness (Jabłoński 2015a). The scalability of the business model of the company embedded in the network and the business model of the network itself means the following:

1. Scalability is expressed by the ability of the business model to continuously adjust its attributes to enable the company to achieve high performance.
2. Scalability is expressed by the ability to adapt the business model to the network business model.
3. Scalability is expressed by the ability of the company to effectively capture value from the network.
4. Scalability is expressed by the ability of the company to continuously effectively use and acquire resources and components of the business model from the network to achieve high performance.

In terms of the configuration, it is important to learn how the business model changes, and which elements have an impact on scalability and why (Juntunen et al. 2018). The concept of business model scalability is limited in this context to the potential of organizational growth or increased turnover to achieve economic benefits (Lund and Nielsen 2018). Business models can be defined as open or closed, which is crucial for assessing their scalability. It is essential to clarify whether their strategies are based on resources and capabilities that cross borders, or on resources and capabilities under the control of the organization. Closed strategies are efficiency-focused and represent the "do-it-yourself" approach, not only relying on resources within an organization but also organizing them to achieve efficiency through standardization, automation and/or centralization. Open strategies, on the other hand, focus on adaptation and are based on the idea of creating together. In both approaches, adaptation strategies focused on new solutions that are based on the use of changing conditions and environments to improve the scalability of the business model are crucial (Hultberg and Pal 2021). As indicated above, the feature describing the adaptability of the business model is its scalability, which contributes to the transformation of more work into even greater efficiency, and consequently, better results (Jabłoński 2016). The patterns that can be used to increase business model scalability were indicated by Nielsen and Lund (Nielsen and Lund 2018) as follows:

- PATTERN A: Adding new distribution channels.
- PATTERN B: Exploring ways to release from traditional capacity constraints.
- PATTERN C: The outsourcing of investments.

- PATTERN D: The leveraging of partners working for free.
- PATTERN E: Implementation of platform models.

These patterns are hybrid, i.e., organizational factors are simultaneously assisted by factors based on digital economy solutions. Figure 2.2 below shows the relationships between different types of scalability.

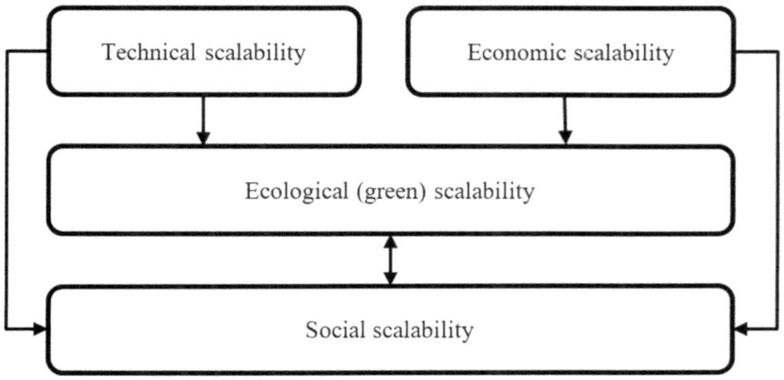

Figure 2.2 Relationships between different types of scalability of business models of companies.
Source: Own study

In this context, taking into account those interdependencies, technical scalability depends on the elements of the technical system together with feedback loops in the context of assigned tasks for specific functions of that system. Economic scalability is conditioned by the relationship between revenue growth rate and the cost structure depending on this process. Social scalability, in turn, determines the dependence of the number of users of the digital business model who affect the social system within the limits of its impact. The scope of impact depends on the power of influence, market share and the market attractiveness of the business model. Green scalability determines the relationship between the number of factors which affect the climate in the context of the environmental impact through industrial and service activities. It covers the impact of business activity outside the created socio-technical system. Companies designed to be digital adapt their business models through an iterative process to achieve scalable operations (Mihailova 2023).

2.5.1 Socio-technical and Cyber-technical Factors of Scalability of Digital Business Models

In the age of digitalization, to better understand management processes, it is necessary to look at business in terms of technical and social criteria.

Technical and social factors in terms of management efficiency must create a coherent system. In ontological terms, the technical system is designed to provide a reliable and predictable link between the user input and the system output. Complex systems are characterized in terms of their scalability by a large number of interdependent elements. A small system is a system consisting of a small (countable) number of elements that can be precisely identified, while a large system is a system with such a large number of elements that it is practically impossible to identify them. In this respect, the number of feedback loops between elements in relation to their possible states, which is a measure of their complexity, is crucial (Mynarski 1981). This relationship should be designed to serve the needs of users. Social systems, on the other hand, are the result of constant evolution and include the processes of change and behavior (Fischer and Herrmann 2011). Socio-technical systems (STS) are dynamic, adaptive systems which derive from the interaction of people who use tools, techniques and knowledge (a technical subsystem) to produce a product or service. This definition takes into account organizations consisting of three interdependent social, technical and environmental subsystems (Di Maio 2014). Socio-technical systems meet the definitions of complex systems, so they require an interdisciplinary approach to solving specific problems (Luther et al. 2023). Technical and social variables together lead to complexity (Schöttl and Lindemann 2015). In this context, business models should therefore be examined as complex socio-technical systems and, if embedded in the digital economy, the digital component should be a key carrier enabling companies to unlock business potential. The relationships between variables such as organizational structure, people, technologies and related tasks are vital. Complexity should express the relationships between technical systems and interactions carried out through tasks performed by people who form communities. Presentation of the complexity of socio-technical system are shown in Figure 2.3.

Social issues are strong determinants of business value. A new theory for the identification of the value-based management framework (Duong Oesterreich et al. 2022) can be seen and created by the socio-technical lens. Value in a competitive economy based on digital transformation processes is created by innovation. In this context, social, behavioral and institutional elements, together with the environment and infrastructure, require an appropriate reconfiguration of the socio-technical system, which can be shaped by the diffusion of many innovations (Hoick et al. 2022). Attention should be paid to the need for automation in various fields, especially taking into account these socio-technical factors (Ibrahim et al. 2021). Therefore, hardware and software technologies of the company system simply manifest

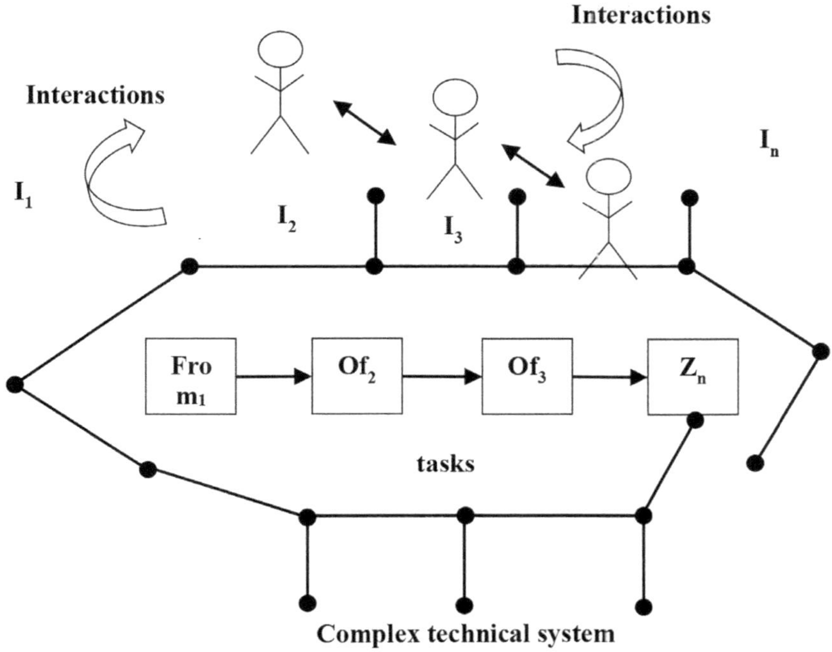

Figure 2.3 Presentation of the complexity of socio-technical systems.
Source: Own study based on Schöttl and Lindemann (2015).

themselves as a complex socio-technical system (Fayoumi and Williams 2021). Artificial intelligence, the role of which in improving business scalability will come to the fore in the near future, will be a factor in this situation, but also a solution that significantly accelerates the processes of achieving the goals of the organization and their projects in the context of innovation determinants. Rather, it is necessary to design the entire work system taking into account socio-technical aspects within a systemic approach which combines factors such as people, organizational issues, technical systems, problems, possibilities and opportunities (Gabriel et al. 2022). A scalable socio-technical system should also demonstrate the characteristics of a resilient system. This is due to the increasing complexity of socio-technical systems. Ensuring their long-term resilience, which will depend on the vitality, flexibility, effectiveness and efficiency of these systems, is therefore justified (Eljaoued et al. 2020). In this context, another element that affects the level of digital business model scalability in the context of the determinants of socio-technical systems, namely the Internet of Things, should be identified. It presents new challenges for experimenting, testing and operating large-scale decentralized socio-technical systems (Fanitabasi et al. 2020). It is important to maintain the expected level of safety as a quality function in the process of changing the scale of work (Foster et al. 2019). In the context of the identification of factors responsible for scalability, the resilience of systems is necessary, which is ensured through cyber-physical solutions. They drive innovation in this way. Cyber-physical systems are a combined hardware and software solution that is deployed in the surrounding physical world and interacts with it. System designers and developers, especially those deployed in a critical environment, often use functional safety activities to assess and manage the risk related to what they create (Rosenthal and Jones 2022). The logic of socio-technical and cyber-physical systems thus understood describes the assumptions for digital business model scalability.

2.5.2 Green Scalability of Digitally Assisted Circular Business Models—Concept Assumptions

Digital business model scalability is a fundamental advantage compared to traditional linear business models. Scalability in technical terms is such a phenomenon where it is possible to expand (increase the scale of operation) the system or undertaking. This is achieved through the relationship between large expenditure on the implementation of the undertaking in the construction of a digital solution in the first phase of implementation and the low costs of their reproduction, sale and distribution. In technical terms, in the case of scalable software, the attribute of this scalability is that business applications can adapt to handling a growing amount of data or a growing number of users. In this situation, scalability is also one of several important attributes of the network in addition to redundancy, security, performance and relatively easy maintenance and management conditions. Hence, the attributes of digital platforms far exceed the possibilities inherent in a standard linear approach to shaping business models along the classic supply chain. As such, it is impossible to implement this scalability. Therefore, competing with companies that have made scalability an attribute of their competitiveness is basically not an option for linear business models. Network scalability is when, without interfering with its logic, it is possible to connect further users to the digital platform. Scalability will therefore have at least two core dimensions, namely technical and economic. The technical dimension will allow the technical system to add more components without having a significant impact on the whole system, which is due to the ability of, e.g., the network to expand. However, it should be remembered that networks based on Internet assumptions are designed precisely in terms of the functionality of scalability, but no system will ever be infinitely scalable. In economic terms, however, the principle is that additional burdens are shifted to partners and users (for example, the Uber digital platform, where additional business components (cars used in the transport process) come from private partners, or the "Too Good To Go" platform, an application that fits into the zero waste trend). It helps combat food waste by stores and restaurants by increasing the global market share by means of the scalability method. Currently it has more than two million users and nearly 200,000 partners who deliver unused meals or food products via a scalable platform. Adding a new user in both the former example and an additional partner in the latter is very simple and low-cost, which fits the assumptions of the aforementioned economic concept of marginal cost, that the cost of adding additional users and partners will be relatively low, or in special cases zero. In addition to the traditional scaling logic centered around growth based on the expansion of a single company and its geographic expansion and turnover, scaling may also involve other, fundamentally different, environmental and social logic. Social impact is decisive, rather than economic incentives. Sustainability-oriented scalability in such terms can be considered far more important than the development of a single company (Sandberg and Hultberg 2021). In this context, the concept of the green scalability of circular business models can be ontologically explained as the use of digitally assisted circular business models where the relationship between the resources involved in the manufacturing process and the ecological effects will be significantly positive. In other words, the resources involved in the manufacturing process and the finished end product will create such residual value that can be reused to the maximum, increasing not only the ecological effect, but also creating new added value. At the same time, it can significantly contribute to the creation of the new space for new circular business models, which create further added value. However, this added value does not have to be a value obtained in a narrow sense by a single company, but can be consumed by the general public, and thus society as a whole.

The concept of the scalability of business models is highly applicable to solutions based on the circular economy assumptions. In order to be effective, circular economy solutions must be scalable (an interesting practical application of circular economy solutions in the context of the need to achieve a high level of scalability is CONTEC, the Polish start-up, which has developed an original method of recovering soot from used tires. It is the only company in the world that uses molten salt as a heat transfer medium in the process of tire pyrolysis. This gives CONTEC an advantage over solutions in which reactors are heated with burners or flue gases. Cf.: Forbes, 2023). Therefore, scalability is a fundamental condition for achieving high efficiency in the implementation of the "Reduce, Reuse and Recycle" strategy, which underlies the circular economy concept. In this situation, scalable solutions guarantee predictability, consistency, efficiency and high-quality inflow of resources (Martina and Oskam 2021). Therefore, in addition to the traditional logic of scaling development-focused companies through the geographical expansion of the company and turnover, scaling could also include a fundamentally different logic in environment and society protection (Sandberg and Hultberg 2021). However, some authors point to the fact that in the case of numerous business models where pollution and resource waste can be reduced, the lack of this scalability is one of the main reasons for the slow transition (Hultberg and Pal 2021). Therefore, the search for effective and efficient mechanisms of the green scalability of business models of companies which apply the assumptions of the circular economy is a key challenge for science from the perspective of developing theories and searching for methods of the application of proposed solutions.

2.6 ORGANIZATIONAL SCALING OF COMPANIES FOR SOCIAL IMPACT

The relationship between the organizational scalability of companies and social impact seems illogical at first glance. However, when both of these concepts are explained unequivocally, doubts already subside. The analysis of the concept of scalability helps define and distinguish, among others, scaling as an organizational process, scalability as a simple organizational ability and scaling up as a phase of organizational development (Coviello et al. 2024). Organizational scaling involves the process of building and strengthening the internal capabilities, systems, and structures necessary to support and sustain growth (Belitski et al. 2023). Scaling a business means increasing the size, scope, and complexity of a business in order to increase its efficiency, improve its productivity, and gain greater market share. This includes the development and implementation of strategies, processes and technologies to enhance the ability of a business to grow and be competitive in the market (Kiradoo 2023). Scalability can be examined in terms of two core areas. The first concerns technical scalability, which is important in terms of socio-technical systems. If we add too many components to the technical system compared to the critical performance of this system, each added component will adversely affect the achievement of the assumed performance and functions of this system. Technical system designers always evaluate systems for their performance limit. If they did not do this, the power, transport and IT systems could fail and lose their critical performance capabilities. In order to prevent such a situation, it is necessary to set a limit level of system components that can work efficiently, creating a comprehensive system. Any additional system component will have an adverse impact on the planned performance of the technical, operating and organizational system. In economic terms, scalability is related to the relationship between revenue growth and cost growth. Such a phenomenon can affect any business model, including digital ones. In the case of the classic approach to business, as a rule, increased revenues are correlated with increased costs. The greater the increase in revenues, the greater costs are. An organization is not economically scalable if the growth rate of sales revenues is slower than the growth rate of sales costs. Such business models are called unscalable. In classical business, everything should be done so that this relationship between revenues and costs is in favor of revenue growth. In the case of flagship business models, where the monetization of business is vitally important, the situation is quite different. In the initial phase of product development, costs increase according to the concept of marginal cost accounting. After reaching the expected standard of a product that meets the expected performance functions, the cost either stops or grows very slowly. Revenues, on the other hand, due to the number of users served, especially for digital platforms, grow exponentially because they are not correlated with further cost increases as they stop at a certain level when the product, digital platform or even service is based on the past cost dedicated in the initial phase of product design and development. At this point, it is necessary to refer to the social impact. Certainly, the optimization of design time, all the more focused on the design of ecological systems and on concern for climate change, is crucial and necessary to achieve progress in the processes of neutralizing the impact of climate change on the quality of life of Earth's citizens. Scalability can be combined with the social impact. If the social impact is positive and works as a simple coupling, beneficial environmental actions translate directly into climate change in the micro version, but if most people in the world apply this solution, the effect of scale will ensure the expected effect. It is worse in a feedback situation when social impact attempts to have a positive impact on the environment, but these activities do not coincide with the opinions of experts and the results of research and analysis. Social impact is therefore crucial in the context of designing scalable business models, including those that are ultimately going to affect climate change. In addition to the economic factor, the social factor also works, where people even demand clean air, clean water, and a pleasant architectural or natural environment. Social impact is not easy to measure because sometimes it is difficult to answer questions about what elements of the earth, and other important elements in these terms, will benefit from this social impact. The scalability of the structure used in the business model can be subjected to scalability in the following formula. The organizational structure is scalable, first of all, if it is flexible and can quickly change its formula without harming employees. Employees can quickly be replaceable, multifunctional, and with interdisciplinary competencies. A flat employment structure, which is the opposite of the hierarchical structure often used in the project orientation of an organization, promotes those project team members who have the greatest technical, communication, social and organizational competences. Then, organizational scalability can underlie the development of new ways of building the competitiveness of companies. Figure 2.4 shows the relationship between the technical scalability and economic scalability of a company's business model.

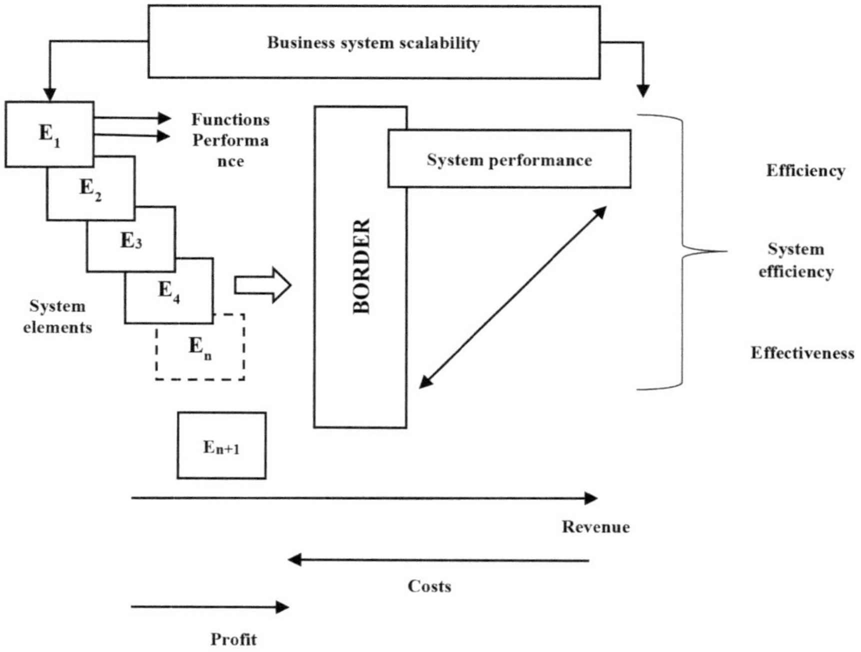

Figure 2.4 The relationship between the technical scalability and economic scalability of a company's business model.
Source: Own study.

The individual elements of the socio-technical system, which constitute a technological business model, enter into a system of interconnectedness aimed at ensuring the efficiency and effectiveness of the designed business solution. Technical efficiency is ensured by the ability of the system to operate even with a large number of elements shaping it, taking into account the threshold number at which the system ensures assumed functions and performance. Economic scalability is maintained when profits are made with specific revenues and costs dependent on the economies of scale of the technological impact of the business model and its interconnectedness with other systems in the context of the delivery and reception of action functions that can also support other business systems through, for example, technology fusions.

2.7 PROPOSAL OF PATTERNS AND THEIR CHARACTERISTICS WHICH DESCRIBE THE GREEN SCALABILITY OF DIGITAL BUSINESS MODELS BASED ON THE ASSUMPTIONS OF THE CIRCULAR ECONOMY CONCEPT

The concept of the circular economy is a contemporary challenge and opportunity for innovative companies. It is based on the assumption that all elements of production and service processes can be included in a circular environment and, by doing so, neutralize the negative impact on the environment, and even in some cases change the climate for the better. The scope of impact and the range of possibilities in this area is unlimited and depends on the creativity and ingenuity of the creators of new technologies and organizational formulas. In general, this concept is based on the transformation of the value chain from a linear approach to a value loop. Table 2.1 presents selected possible directions of applications and solutions of the circular economy.

Table 2.1 Selected Circular Economy Solutions used in the Reconfiguration of Business Models of Companies

No.	Determinant	Explanation
1.	Reuse	Products are repeatedly reused by different end-users during their lifetime.
2.	Recondition	Repair of the defect/improved aesthetics of the product, but without a new/additional warranty on the product as a whole. This includes areas such as: • Repair—removing defects or damage that restores the product to the required efficiency, • Renovation—restoring the required aesthetic parameters to the product.
3.	Remake	Regeneration – includes production series and stages that work on an end-of-life part or product to restore them as new or improve performance, with adequate warranty.

Contd.

No.	Determinant	Explanation
4.	Circular supply	Sourcing recycled materials or renewable materials that can be returned to any of them – a technical or biological cycle. It contains: • Materials extracted from the product at the end of the shelf life instead of using primary materials extracted from finite sources (a technical cycle). • Use of renewable materials/biomaterials that can be restored to the biological cycle.
5.	Recovery of by-products	Residual or secondary products from one process (or value chain) become inputs for another process (or value chain).
6.	Direct use/ industrial symbiosis	Residual products or by-products from one process become a raw material/input into another process. Proximity is usually a key enabler of this mechanism, as transport costs, environmental impact and distance between facilities would not allow this.
7.	Indirect use	Residual products (which may otherwise be considered "waste") gain value as a commodity/raw material for another process (e.g., fly ash from coal, combustion used as cement clinker). They can be combined and sold through the secondary market for these materials.
8.	Recovery of resources	End-of-life materials or products are recovered and reintroduced as inputs into the value chain. The categories include: • **Circular recycling:** End-of-life materials are recovered for reuse in the same value chain. Importantly, the integrity of materials is maintained, allowing them multiple or even infinite cycles. • **Open loop recycling:** Materials at the end of the product life cycle have a different value in the chain. • **Upcycling:** The material or product at the end of its shelf life is used as a starting material for production, e.g., as a higher-value product (e.g., in fashion or additives). • **Downcycling:** The material or product at the end of its shelf life is used as a starting material for the production of a lower-value product (e.g., processing of building materials or other products into road aggregates).
9.	Waste-to-energy conversion	Often presented in the context of the circular economy. Waste is converted into energy when materials or products are in an end-of-life process and, for example, are incinerated or processed (e.g., pyrolysis, biofermentation) to produce heat, electricity or fuel.

Source: Own study based on Smith-Gillespie (2020).

Based on the possibilities of application of this approach, patterns for the green scalability processes of digital circular business models have been designed.

They are as follows:

- **PATTERN A:** Develop technological solutions where the level of reduction, recovery, recycling, replacement and renewal generates the highest yield.
- **PATTERN B:** Establish mechanisms for eliminating technological, operational and organizational constraints in the process of value loop implementation.
- **PATTERN C:** Ensure such a relationship between the expenditure on and outcome of the implementation of the circular economy solution that, through the volume scale of the processed resource, the yield is the largest.
- **PATTERN D:** Ensure that the social and environmental impact during the implementation of the circular economy solution is significant.
- **PATTERN E:** Ensure that the participation of digital platforms in the control, monitoring and generation of yield from primary resources is an essential condition for implementing these solutions.
- **PATTERN F:** Maximize the level of revenue generated from circular economy solutions.

The six patterns developed for the green scalability processes of digital circular business models define a comprehensive set of recommendations, the implementation of which should contribute to the performance of a set of activities by a single company. This, in turn, should create a partial value which contributes to positive climate change, inter alia, by changing the structure of the production processes of products and the provision of services, allowing them to be optimized by creating new revenue opportunities based on resources already used. In this way, the global demand for new products should decrease and the level of substitution thereof by the new forms of transformation of available resources should increase significantly.

The author's original name and concept of the green scalability of digitally assisted circular business models were proposed. It was shown that the proposed concept of the green scalability of circular business models shifts the focus from the scalability of a single company to scalability, the effects of which will benefit not only the company itself, but also a wider range of recipients—society. Scalability, thus defined, increases the positive social impact of business.

2.8 CONCLUSIONS

The content presented in the chapter showed how important the concept of scalability is for the design of digital business models, and what types of scalability can occur in the process of searching for optimal solutions in the sphere of business ecosystems. Interdependence between ecosystems, adding new users and thus creating new interactions between them requires the provision of such technologies where, if new users join the system, the system will continue to perform its functions. At the same time, it will not be limited by poorly selected technologies. The cost parameter significantly affects the technological scalability in question. In turn, economic scalability is a condition for achieving profits through the scale of digital business models. Taking into account the scale of freedom in the processes of designing and developing digital business models is necessary to understand this issue fully and achieve the expected results in business.

REFERENCES

Belitski, M., Martin, J., Stettler, T. and Wales W. 2023. Organizational scaling: The role of knowledge spillovers in driving multinational enterprise persistent rapid growth. Journal of World Business. 58: 2.

Cavallo, A., Cosenz, F. and Noto, G. 2023. Business model scaling and growth hacking in digital entrepreneurship. Journal of Small Business Management. DOI: 10.1080/00472778.2023.2195463.

Coviello, N., Autio, E., Nambisan, S., Patzelt, H. and Thomas, L.D.W. 2024. Organizational scaling, scalability, and scale-up: Definitional harmonization and a research agenda. Journal of Business Venturing. 39: 4.

Di Maio, P. 2014. Towards a metamodel to support the joint optimization of socio technical systems. Systems. 2(3): 273–296.

Duong Oesterreich, T., Anton, E., Teuteberg, F. and K Dwivedi, Y. 2022. The role of the social and technical factors in creating business value from big data analytics: A meta-analysis. Journal of Business Research. 153: 128–149.

Eljaoued, W., Ben Yahia, N. and Bellamine Ben Saoud, N. 2020. A qualitative-quantitative resilience assessment approach for socio-technical systems. Procedia Computer Science. 176: 2625–2634.

Espina-Romero, L.C., Guerrero-Alcedo, J.M. and Ossio, C. 2023. 7 topics that business ecosystems navigate: Assessment of scientific activity and future research agenda. Heliyon. 9.

ESSEC Chair of Networked Business Models. 2017. Business ecosystems: A literature review. Working Paper, Economics of Mutuality Alliance.

Fanitabasi, F., Gaere, E. and Pournaras, E. 2020. A self-integration testbed for decentralized socio-technical systems. Future Generation Computer Systems. 113: 541–555.

Fayoumi, A. and Williams, R. 2021. An integrated socio-technical enterprise modelling: A scenario of healthcare system analysis and design. Journal of Industrial Information Integration. 23: 100221.

Fischer, G. and Herrmann, T. 2011. Socio-technical systems: A meta-design perspective. International Journal of Sociotechnology and Knowledge Development. 3(1): 1–33.

Foster, C.J., Plant, K.L. and Stanton, N.A. 2019. Adaptation as a source of safety in complex socio-technical systems: A literature review and model development. Safety Science. 118: 617–631.

Gabriel, S., Bentler, D., Grote, E.-M., Junker, C., Meyer zu Wendischhoff, D., Bansmann, M., et al. 2022. Requirements analysis for an intelligent workforce planning system: a socio-technical approach to design AI-based systems. Procedia CIRP. 109: 431–436.

Gao, R. 2021. The theory and development of business ecosystems. Advances in Economics, Business and Management Research. Volume 203.

Hallowell, R. 2001. Scalability: The paradox of human resources in e-commerce. International Journal of Service Industry Management. 12(1): 34–43.

Hill, M.D. 1990. What is scalability? ACM SIGARCH Computer Architecture News. 18(4): 18–21.

Hoicka, C.E., Zhao, Y., McMaster, M.-L. and Das, R.R. 2022. Diffusion of demand-side low-carbon innovations and socio-technical energy system change. Renewable and Sustainable Energy Transition. 2: 100C34.

Hultberg, E. and Pal, R. 2021. Lessons on business model scalability for circular economy in the fashion retail value chain: Towards a conceptual model. Sustainable Production and Consumption. 28:686–698.

Ibrahim, A., Kyriakopoulos, S. and Pretschner, A. 2021. Causality-based accountability mechanisms for socio-technical systems. Journal of Responsible Technology. 7–8: 100016.

Jabłoński, A. 2015a. Skalowalność modeli biznesu w środowisku sieciowym [Business model scalability in the network environment]. CeDeWu, Warszawa.

Jabłoński, M. 2015b. Ocena atrakcyjności inwestycyjnej modeli biznesu [Assessment of the investment attractiveness of business models]. Zeszyty Naukowe Uniwersytetu Szczecińskiego nr 854, Finanse, Rynki Finansowe, Ubezpieczenia. nr 73: 455–468.

Jabłoński, A. 2016. Scalability of sustainable business models in hybrid organizations. Sustainability. 8(194) doi:10.3390/su8030194.

Juntunen, M., Ahokangas, P. and Nguyen, H. 2018. Business model scalability in the cloud business context. Journal of Business Models. 6(1): 19–39.

Kim, H., Nam Lee, J. and Han, J. 2010. The Role of IT in Business Ecosystems, Communications of the ACM. 53(5).

Kiradoo, G. 2023. Scaling your business: Growing your impact and reaching your goals. Current Topics on Business, Economics and Finance. 2: 197–224 Available at SSRN.

Lund, M. and Nielsen, Ch. 2018. The concept of business model scalability. Journal of Business Models. 6(1): 1–18.

Luther, B., Gunawan, I. and Nguyen, N. 2023. Identifying effective risk management frameworks for complex socio-technical systems. Safety Science. 158: 105989.

Martina, R.A. and Oskam, I.F. 2021. Practical guidelines for designing recycling, collaborative, and scalable business models: A case study of reusing textile fibers into biocomposite products. Journal of Cleaner Production. 318: 128542.

Mihailova, I. 2023. Business model adaptation for realized international scaling of born digitals. Journal of World Business. 58(101418).

Moore, J.F. 1993. Predators and Prey: A new ecology of competition. Harvard Business Review. 71(3): 75–86.

Moore, J.F. 1996. The Death of Competition: Leadership and Strategy in the Age of Business Ecosystems. New York: HarperCollins.

Moore, J.F. 2006. Business ecosystems and the view from the firm. The Antitrust Bulletin. 51(1): 31–75.

Mynarski, S. 1981. Elementy teorii systemów i cybernetyki [Elements of Systems Theory and Cybernetics], PWN, Warszawa.

Nielsen, Ch. and Lund, M. 2018. Building scalable business models. MIT Sloan Management Review.

Noviyanti, N. and Tricahyono, D. 2024. Business ecosystem development in corporate business strategy. Journal of Multidisciplinary Academic Business Studies (JoMABS). 1(4): 645–662.

Peltoniemi, M. and Vuori, E.K. 2004. Business ecosystem as the new approach to complex adaptive business environments. *In*: M. Seppä, M. Hannula, A. M. Järvelin., Kujala J., Ruohonen M., and T. Tiainen, Frontiers of E-Business Research. 267–281. Tampere: University of Tampere.

Rong, K., Lin, Y., Li, B., Burström, T., Butel, L. and Yu, J. 2018. Business ecosystem research agenda: more dynamic, more embedded, and more internationalized. Asian Bus Manage. 17: 167–182.

Rosenthal, C. and Jones, N. 2022. Inżynieria chaosu, odporność systemów w praktyce [Chaos Engineering. System Resiliency in Practice]. Hellion S.A., Gliwice.

Sandberg, E. and Hultberg, E. 2021. Dynamic capabilities for the scaling of circular business model initiatives in the fashion industry. Journal of Cleaner Production. 320: 128831.

Schöttl, F. and Lindemann, U. 2015. Quantifying the complexity of socio-technical systems—A generic, interdisciplinary approach. Procedia Computer Science. 44: 1–10.

Smith-Gillespie, A. 2020. Defining the concept of circular economy business model. The route to circular economy. Project funding from the European Union's Horizon 2020 Research and Innovation Programme under Grant Agreement No. 730378.

Stampfl, G., Prügl, R. and Osterloh, V. 2013. An explorative model of business model scalability. International Journal Product Development. 18(3/4).

Chapter **3**

Socio-technical Systems as a Tool for Creating a Platform for Innovative Business Models

3.1 INTRODUCTION

Embedding socio-technical systems in modern business is well-established, but as regards the digital economy, it has not been examined as widely so far. The new determinants of innovative business solutions require that this issue be looked at differently, that is, taking into account technological aspects in terms of the conditions of digital technology and the transformation of processes towards business digitalization. The network paradigm is particularly important in this regard, which, together with the system paradigm, explains the principles of social communication, i.e., in this situation business communication as well, taking into account the fact that a large group of international recipients is centered around digital platforms, which significantly affects the formation of cultures, behavior and creation of value for shareholders. Social enterprises also play an important role, as they can achieve their social economic goals faster and more efficiently thanks to business ecosystems and faster access to potential stakeholders. This chapter focuses on these issues.

DOI: 10.1201/9781003456421-3

3.2 DETERMINANTS OF THE CONSTRUCTION OF SOCIO-TECHNICAL SYSTEMS IN MODERN BUSINESS

The core determinant of the construction of socio-technical systems is connecting the human community with technological factors. The modern challenges of business digitalization mean that a number of technological solutions emerge around people's lives that fit into the context of people's functioning in the professional and life ecosystem. Business leaders and their organizations will also need to acquire more knowledge about the evolution of the technologies, understand the art of opportunity, and the potential for the future to best prepare their businesses for the use of automation (Wilburn and Wilburn 2018). The importance of technology in the business sector has been strongly recognized. It has become a huge business in its own right, in the form of software design and development, hardware manufacturing and robotics applications (Santhosh 2018). Recognizing this aspect in the context of socio-technical systems theory set against the background of commonly used digital platforms is a core element of research in the area of explaining the determinants responsible for shaping effective and efficient business ecosystems. In this environment, business models are created that are based on the assumptions of the concept of socio-technical systems. Innovative solutions for technology are crucial, so that they can be used in companies in both their processes and in their products, to enable a competitive advantage (Saura et al. 2022). Nowadays, in the era of digital transformation, innovations built from socio-technical systems have been appreciated and their achievements are useful in designing new business solutions. The complexity of technology-based business systems is a crucial aspect while examining the socio-technical system. Complex systems such as socio-technical systems require detailed research, analysis and the assessment of their applicability in the business environment. Man versus machine, from the cybernetic perspective, is a leading topic at the intersection of management, engineering and sociological research. The issue of explaining the place and role of socio-technical systems in the design of digital business models is interdisciplinary. Theoretical aspects of understanding socio-technical systems in the context of designing business models based on the complexity of systems require the identification of features characterizing these systems in relation to other approaches to understanding business ecosystems of the modern economy. The digital transformation makes it necessary to consider the management perspective against the background of the development of innovative technologies supporting business. In order to take the lead in fierce market competition, companies need to pay attention to the application of technological innovation and increase the level of scientific and technological innovation (Ni 2018). This approach deviates significantly from the traditional understanding of management-based business, including decision making based on traditional management structures. In the case of digital platforms based on complex socio-technical systems, many functions are performed automatically, whereby a person as an operator or user navigates a pre-designed algorithm of conduct. Algorithms of conduct that combine people's needs, their preferences and predictions for future needs mean this basis of the functioning of modern human society depends to a very large extent on mass-implemented technological solutions. They underlie people's behavior, their communication and how they do business. Such an approach requires synergy between what is human and what results directly from the functionality of technical systems.

3.3 SOCIAL BUSINESS ECOSYSTEMS

A business ecosystem is a set of all their entities and their activities in a specific group of companies, usually a sector or industry focused on the same objectives and operating in a specific geographical area. Social entrepreneurs try to grow businesses with little investment and limited resources. Interactions including networks between social entrepreneurs, investors and other stakeholders are essential to promoting social entrepreneurship (Cato and Nakamura 2022). A different link between the two areas emerges, not only between the socio-technical systems, but between the social and ecological systems. Socio-ecological systems, known as the "complex man–earth system" or the complex man–nature system, refer to a system coupled with the complexity, non-linearity, uncertainty and multi-layered embeddedness features created by human–environment interaction (Liu et al. 2023). The social aspect is broader than the ecological one, but the latter is contained therein. However, the research results described in the relevant literature show that research mainly focuses on antecedents, the establishment and management of social enterprises, but less attention is paid to sociological perspectives, such as the theory of social management and population, and there are no conclusions on entrepreneurial performance (Wang and Mui 2023). An ecosystem can also include an IT environment based on socio-technical systems, taking into account the crucial role of people as system operators, but also people as users. The social and economic areas play a leading role within the business ecosystem. They determine the attractiveness of the business ecosystem and therefore the dynamics of functioning in this ecosystem. The greater the investment attractiveness of the business ecosystem, the greater the share of business units and stakeholders involved. In the digital economy, digital platforms shape the relationships between entities communicating through these platforms. Platforms foster relationships and facilitate and control cash flows. The social factor, especially in the case of a large number of community users, determines the intensity of the relationship, playing a fundamental role. The social ecosystem of business is one where not only the interests of the individual are assessed, but in particular the interest of the general public. The digital platform is intended not only to serve the needs of these individuals, but to provide added value to a large group of users and stakeholders. The key objective of the functioning of the social business ecosystem is to create a positive social impact within the ecosystem and beyond its borders. The business ecosystem in this sense will be based on the theoretical assumptions of the concept of socio-technical systems. Technologies, infrastructures and their interfaces, as well as people and their relationships, should shape social and technical culture in a synergistic way. In this way, an ecosystem of modern business is built, where the division of individual roles in the value chain is based on the specialization and delegation of powers to specific levels of management. Figure 3.1 shows the relationships between the creation of the social ecosystem of business based on the assumptions of the functioning of social enterprises.

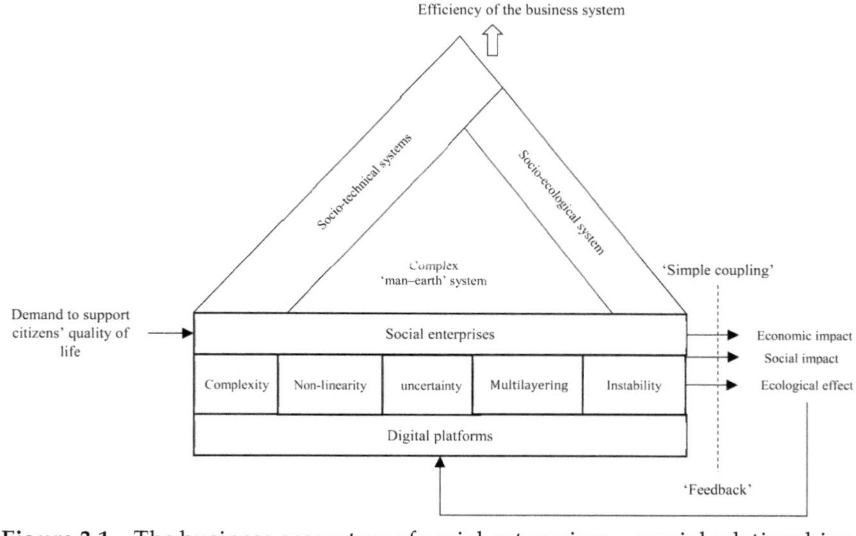

Figure 3.1 The business ecosystem of social enterprises – crucial relationships. *Source:* Own study.

The efficiency of the business ecosystem depends on the relationship between socio-technical and socio-ecological systems as regards the functioning of social enterprises. Digital platforms perform economic and social functions by concentrating large numbers of users and creating space for the exchange of economic and social values. From this perspective, economic, social and ecological impact is created, implemented by means of simple couplings in terms of the theory of cybernetics. In turn, feedback loops are achieved through these types of impact on the improvement of the functioning of digital platforms which underlie the design of digital business models.

3.4 SOCIAL ASPECTS OF BUSINESS MODELS

Business models focus on capturing value from the market. Business model configuration, income generation logic, customer service, cost structure and an income model are just a few of the core variables that describe business models. At the same time, business models based on social entrepreneurship are created. Models are created to effectively meet basic human needs that current markets and institutions have failed to meet. Social entrepreneurship combines the ingenuity of traditional entrepreneurship with the mission of changing society (Seelosa and Mair 2005). The social aspects of business models are focused on the following issues:

- Maximizing the volume of recipients and recommending users and partners of the business model, which is related to the scale of the impact of the business model on the social environment.
- Maximizing the created value in the context of the evidence of social impact achieved by the core tasks of the business model.
- Creating a positive environmental impact through smart circular economy solutions.

In theoretical terms, several typologies of business models implementing social undertakings can be identified. For example, such a typology was proposed by L. De Cuyper, B. Clarysse and M. Wright, who defined models such as social stimulators, social providers, social producers and social intermediaries. As regards value proposition and the symbolic use of social values, social simulators focus on social value proposition (social values conveyed in products/services sold), while social providers concentrate on functional value proposition (the functionality of a product but possibly adapted to the needs of the customer–beneficiary). As a result, social enterprises that collectively pursue commercial and social value are gaining increased popularity

(Cuyper et al. 2024). Social factors, including ecological ones, are crucial for business solutions that use popular ideas acceptable to society. Companies therefore innovate in their business models to deal with the economic, environmental and social aspects of sustainability and their interrelationships (López-Nicolás et al. 2021). Climate change and social expectations arising from an increased awareness of environmental responsibility force business model designers to add components related to intervention towards changes in business perception—from a focus on profit to a focus on social utility, understood as providing added value to the company through added value to society. Thus, in addition to the value for investors, the benefits from the activities of companies are received by the general public, who receive factors that improve the quality of life of citizens. Then, acceptance of the functioning of business is greater, because there are many more beneficiaries as key stakeholders of companies. However, in this value creation process, social enterprises and conventional for-profit organizations differ in whether they treat economic value and social benefits as "means" or "objectives" of their businesses (Zeyen et al. 2014). At the same time, a balance of influence is necessary in order to preserve the rationality of managing scarce resources. The role of business models focused on positive climate change related to weather monitoring services, information systems, drainage solutions, water management, and energy management including energy clusters, etc. is also growing.

3.5 SOCIAL COMMUNITY BUSINESS MODELS WITH A POSITIVE IMPACT ON THE ECONOMY

Building large communities focused on consuming the value created by the logic of a specific business model underlies the design of modern digital platforms. These mobile-assisted platforms are shaping innovative business ecosystems, based on the assumptions of socio-technical systems. Digital technology is a condition for merging parties for commercial transactions. To this end, the concept of the sharing economy is used, where resources are shared by different people. This works on the principle of "I have, you need", so resources unused by individuals can serve at a time when they are not needed by other people. In this way, the ecological impact is achieved and the community benefits from these goods. The paradigm of ownership is changing in favor of temporary use. A social, moral form of action is the decision of managers and owners to act for the common good (de la Cruz Jara and Spanjol 2024). This approach demonstrates a very high degree of public awareness of how to optimize the use of scarce resources. This approach is particularly popular among Generation Z, i.e., the generation of post-millennials, for whom mobile phones are a core element of their lives. Communities revolve around ideas that typically address aspects such as protection from climate change, equality, human rights, accessibility and a lack of desire to own in favor of use. Community building is a primary business objective in this context, while commercial transactions are secondary, as having a large community volume increases the likelihood that some of them will spend their funds and thus monetize the business model. As a result, there is synergy between the benefits for users, which are economically acceptable to them and the owners of digital platforms who can generate profits at the same time. The condition for users to accept that the platform generates profits is that these benefits for users allow them to benefit from goods which they would have to pay for in the old approach to the economy. Sustainable businesses are created taking into account social, economic and technological factors (Vargas et al. 2024). Such a crowd-oriented business provides a better quality of life for new generations of citizens. The condition for achieving social impact is therefore the mass of people grouped around a digital platform based on a specific idea, who are not only customers, users, but in particular ambassadors of the solution used. Digital platforms meet the definition of socio-technical systems because, on the one hand, they provide a social factor based on the positive acceptance of the value received and, on the other hand, technology makes it possible to build such a business model. In order to better explain the place and role of social communication in the context of the determinants of the functioning of digital platforms, a triad of the construction of this social communication has been developed, as shown in Figure 3.2.

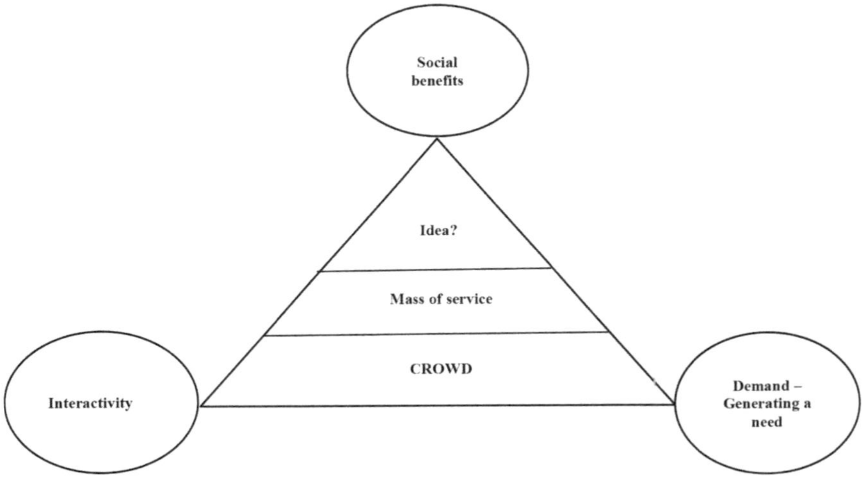

Figure 3.2 The triad of building social communication through digital platforms. *Source:* Own study.

The triad is based on three fundamental assumptions about building relationships between the users of digital platforms focused on creating social values. These include interactivity, demand—generating a need—and social benefits. In these three dimensions, the crowd is focused on using the platform for their needs and, at the same time, meeting new needs in social and ecological spheres. This requires the digital mechanisms of mass user service and an attractive and feasible idea. The attractiveness of such social business models depends on the ability to respond to the current needs of increasingly vulnerable societies.

3.6 A NETWORK PARADIGM AND SOCIO-TECHNICAL SYSTEMS

The network paradigm contrasts with the system paradigm. Its features seem different from the system paradigm, as the focus is on relationships, nodes and patterns of connections. Although the network is also a system, these paradigms are clearly distinguished in the relevant literature. The network is formed by smaller groups of relationships combined into a whole. The network is based on communication between network members in the context of specific nodes. Network members are called actors. The actor is usually seen as a rational, active agent who uses their position in the network to maximize profit. The actor's position in the network is described in terms of a desired abstract pattern of connections, such as having a rare ego network or being located along the shortest path between otherwise unrelated actors. The benefits to the actor are mainly a function of the topology of the local network, and associations are seen by default as forming a leverage structure (Borgatti and Foster 2003). The effectiveness and efficiency of marketing strategies and plans can be improved through network relationships (Piercy and Cravens 1995). A network is a socio-technical system if it is based on technological solutions usually related to the digitalization of industry and business based on digital platforms. The place and role of humans in the system of relationships without technological solutions resulting directly from digitalization would not take place on such a scale that one can talk about the network. In such a situation, sociometry makes it possible to measure network indicators such as the number of ties, orientation, their reciprocity, transitivity, centrality, density, strength, bridges, brokerage and equivalence (Prell 2011). The identification of these parameters helps explain the attributes of the network and their impact on the internal and external environment. Shaping the network allows for the creation of a business ecosystem based on relationships and the intensity of contacts of the community forming the network. Often, people-initiated online activities can be carried out without their direct participation, albeit under their supervision. Modern networks use mass data sets, where data from network activity is collected. The network is, as it were, a social ecosystem that creates specific conditions for the activity of network participants. It is important to be aware of the quality of the network members' role in its development and impact on operational activities. The strength of the network is the strength and intensity of the relationship between its members. These relationships can be positive, but in some cases are also negative, which can be a problem in terms of creating value from online activity. Therefore, the ethics and principles of cooperation, which should be established and respected, are crucial. Software for network development based on the construction of a socio-technical system should be used to control processes in the network. If the processes in the network are well understood, and network coordinators are aware of network attributes, then such a solution can be considered to serve to build an environment where economic entities can exist and develop. A network based on a socio-technical system is currently one of the most popular business creation environments based on the digital transformation of the world economy. Figure 3.3 shows the relationship between the network paradigm and the socio-technical system.

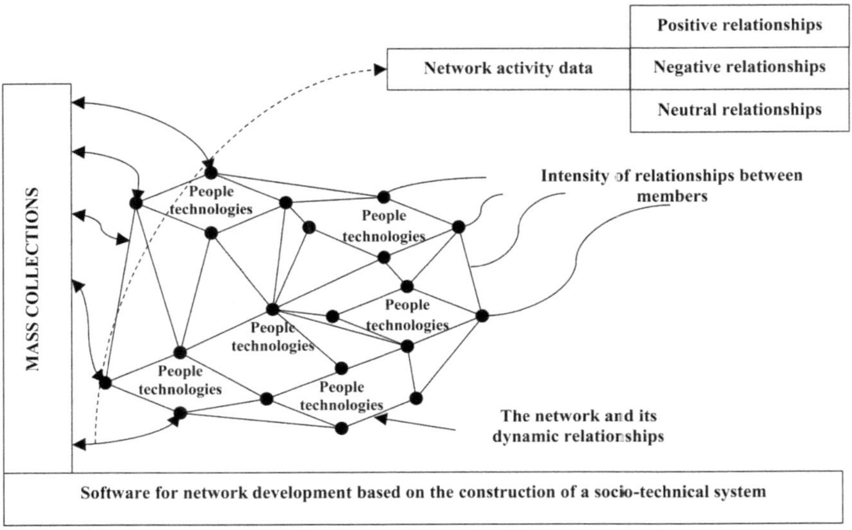

Figure 3.3 The relationship between the network paradigm and the socio-technical system. *Source*: Own study

The business network is based on two core variables, i.e., actors (people) and technologies enabling the integration of these people with each other. The business effect of the network is the degree of relationships between its participants, which can be positive, negative or neutral. The use of mass data stimulates network relationships and enables the achievement of network goals and the goals of individual network actors. The network creates business models based on the potential of the network. The better the effects of networking, the greater the chance of creating innovative business models.

3.7 SOCIAL ENTERPRISES IN DIGITAL BUSINESS ECOSYSTEMS

Social enterprises are enterprises whose mission is not only to generate profits, but in particular to create social impact. In their statutes, these companies declare to act for the social good based on the adopted business model. Social enterprises (SE) have recently gained more attention and have been widely recognized as a viable alternative method to promote social well-being. They have also become one of the core instruments for promoting inclusive growth in a country, with a particular focus on people and social cohesion (Rizal Palil et al. 2023). Social enterprises are focused on reversing imbalances in social, structural and political systems by creating and sustaining positive social changes that may be the product of religious impulses, social movements, cultural or professional interests, feelings of solidarity and reciprocity, altruism and, more recently, the government's need to assist in the performance of public functions (Trivedi and Stokols 2011). A social enterprise is a separate combination of social and environmental missions with income-generating activities (Vicente 2022). The business model of a social enterprise focuses on such a configuration of its components that are subordinated to the implementation of the social mission. The usefulness of products or services must first meet the needs of society, specific individuals, social and ecological responsibility, as well as fitting into the modern trends of the ESG concept (Environmental, Social and Governance). Today, social enterprises are no longer a much smaller share of companies in general, but rather are growing in number due to the growing awareness of citizens about environmental responsibility, concern for climate change, the quality of life of people as well as equality and respect for human rights. Social enterprises or social entrepreneurship have been conceived as focusing on creating value for the good of society or the environment, rather than capturing the value typical of commercial enterprises (Lipei 2017).

The business model of a social enterprise does not differ particularly from a business model focused on commercial activities; however, all components of this business model are focused on and sensitive to social impact, taking into account the adopted model of a non-profit organization. The social enterprise ecosystem is based on cooperation and creates added value between people; hence, the essence of its determinant is the number of stakeholders gathered around a specific supportive idea that builds social value. Triad of social enterprise functioning as a digital platform are shown in Figure 3.4.

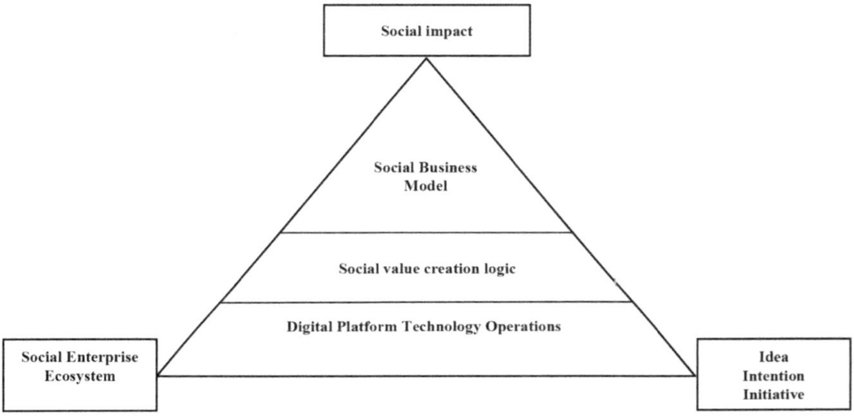

Figure 3.4 Triad of social enterprise functioning as a digital platform.
Source: Own study.

In theoretical terms, theories inherent in the design of social enterprises include thematic areas, concepts, methods and approaches such as the theory of planned behavior, the model of entrepreneurial intention, the modified Hocker's model, social capital theory, social network theory, institutional theory, event system theory, the performance measurement model, etc. (Wang and Mui Yee 2023). In principle, there are no restrictions on the operating space of this type of company. In every sector of the economy, the creation of social value while creating economic value is possible. Well-managed social enterprises can strike a balance between economic efficiency and social welfare. This is what distinguishes them. In the digital economy, the creation of social enterprises is easier because the business model of such an enterprise can function as a digital platform inscribed in the assumptions of concepts such as the sharing economy, the circular economy or Big Data. Due to its space of influence, the digital economy—far larger than traditional businesses—can bring together a huge number of stakeholders, users or fans, who can successfully distribute social value, taking advantage of the opportunities provided by these business models. The digital business ecosystem in this situation contributes to the dissemination of popular innovative social ideas without which one cannot talk about progress in the field of the value economy, focused on creating value for society. Therefore, the balance between what should be effective and what should have a social impact is crucial, and all this should take place in a digital business ecosystem focused on the mass processing of data aimed at increasing the well-being of the population.

In order to explain the functioning of social enterprises in the form of digital platforms, a triad of social enterprises functioning as a digital platform has been developed. It consists of three core elements, namely, firstly the need to build a social enterprise ecosystem that brings the community together around a supportive idea; secondly, the definition of the underlying ideas, intentions and social initiatives that a given business model will implement; and thirdly, the implementation of the expected social impact. These three core elements should be implemented by means of the manner of operationalizing the actions. For the purposes of this operationalization, it is necessary to choose a specific technology for the operation of the digital platform, to develop assumptions for the logic of creating social value, and consequently to adopt a social business model based on the operation of the digital platform.

3.8 CONCLUSIONS

The analysis, discussion and interpretations in the chapter proved that the social factor in the context of the determinants of the design and development of digital platforms is fundamental to achieving their functionality and expanding the range of activity by looking for a mass audience who are interested in creating a community that creates value not only for themselves, but also for the external environment, creating a positive social and ecological impact. Social communication, the impact of networking, social aspects of business models and social aspects of business ecosystems allow for the creation of a new perspective, which is balanced compared to the needs of customers and the pressures and expectations of stakeholders.

REFERENCES

Borgatti, S.P. and Foster, P.C. 2003. The network paradigm in organizational research: A Review and Typology. Journal of Management. 29(6): 1004.

Cato, S. and Nakamura, H. 2022. Understanding the function of a social business ecosystem. Sustainability. 1.

Cuyper, L.D., Clarysse, B. and Wright, M. 2024. Doing good while making profits: A typology of business models for social ventures. Strategic Entrepreneurship Journal. 273.

de la Cruz Jara, M.F., Spanjol, J. 2024. Understanding multiple perspectives on social value in business: An integrative review and typology. Journal of Business Ethics.

Lipei, H. 2017. The impacts of social enterprise from an organizational perspective: A case study in China, Advances in Social Science. Education and Humanities Research. 119: 1821.

Liu, F., Dai, E. and Yin, J. 2023. A review of social–ecological system research and geographical applications. Sustainability. 15: 2.

López-Nicolás, C., Nicolás, J.R. and Mateo-Ortuño, E. 2021. Towards sustainable innovative business models. Sustainability. 13: 1.

Ni, H. 2018. Study on the role of technological innovation in business administration. Modern Economy. 9: 1619.

Piercy, N.F. and Cravens, D.W. 1995. The network paradigm and the marketing organization. European Journal of Marketing. 29(3): 7.

Prell, Ch. 2011. Social Network Analysis: History, Theory and Methodology. Sage.

Rizal Palil, M., Syuhada Abu Hassan, N., Aaina Aqilah Jamaluddin, N., Suriya Ismail, I. and Ervina Samsuddin, M. 2023. Social Enterprises Entrepreneur and Social Impacts to Community: A Review Paper. Journal of Social Sciences and Humanities. 45.

Santhosh, T. 2018. Impact of technology in business. International Journal of Management and Humanities (IJMH). 3(6): 9.

Saura, R.J., Skare, M. and Riberio-Navarrete, S. 2022. How does technology enable competitive advantage? Reviewing State of the art and outlining future directions. Journal of Competitiveness. 14(4): 180.

Seelosa, C. and Mair J. 2005. Social entrepreneurship: Creating new business models to serve the poor. Business Horizons. 48: 241.

Trivedi, C. and Stokols, D. 2011. Social enterprises and corporate enterprises: Fundamental differences and defining features. Journal of Entrepreneurship. 20(1): 4.

Vargas, J.C., Giner, G. and Clemente, I. 2024. Effect of the economic, social and technological factors on sustainable entrepreneurship over time. Journal of Business Research. 173: 1.

Vicente, L. 2022. The social enterprise: A new form of enterprise? The American Journal of Comparative Law. 70(1): 155.

Wang, A. and Mui Yee, C. 2023. A literature review of social entrepreneurship. Open Journal of Business and Management. 11: 2232.

Wilburn, K.M. and Ralph Wilburn, H. 2018. The impact of technology on business and society. Global Journal of Business Research. 12(1): 34.

Zeyen, A., Beckmann, M. and Akhavan, R. 2014. Social entrepreneurship business models: managing innovation for social and economic value creation. pp. 107–132. *In*: C.P. Zimth and C.G. von Mueller (eds). Managementperspektiven für die Zivilgesellschaft des 21. Jahrhunderts: Management als Liberal Art. Wiesbaden: Springer-Verlag.

Chapter **4**

Socio-technical Systems as a Platform Ecosystem

4.1 INTRODUCTION

Technical aspects in the digital economy are the foundation for creating innovations based on Schumpeter's creative destruction. Significant technological progress, changing sectors and introducing so-called game changers, is changing the conditions of the game in sectors. In principle, in every area of life, technology changes the principles of the functioning of basic functions that humans deal with, and in particular, changes the principles of running a business. Solutions built on digital platforms, often supported by artificial intelligence algorithms, are revolutionizing ways of delivering value, serving customers and manufacturing products. It is important to ensure the level of process security and cybersecurity. Technical ecosystems interact with each other based on interfaces that should be clear and secure, which is a condition for the proper functioning of electronic systems. These are the issues addressed in this chapter.

4.2 A SOCIO-TECHNICAL PERSPECTIVE ON THE DIGITAL ERA

Socio-technical systems are particularly important and are used as a theoretical premise for application in the digital economy era. Theoretical potential is based on the assumption of a combination of social factors that take into account the role of humanism in the design of organizations

DOI: 10.1201/9781003456421-4

and the potential embedded in innovative technologies, which are conducive to the possibility of describing digital business models. The dynamically developing digital economy dynamizes activities aimed at designing theoretical assumptions that are consequently used after verification and validation for application solutions. The combined implementation of theory and practice allows for the formation of new forms of value creation for customers. Software and hardware used in the context of consistent and configuration-based elements which underlie the construction of business models connected by interfaces create the foundations for the standardization of designed solutions. In this way, scalable solutions are created that enable the financial monetization of business solutions. Such solutions are also expected by the market, where the achieved business effectiveness and efficiency as well as the scale of market impact proves the success or failure of the business model. Socio-technical systems understood in this way, supported by innovative technologies, enable the creation of new spaces in the field of business value. This value is embedded in the configuration of the component system, while at the same time creating formulas for the delivery of products and/or services that are much more efficient than in the classical economy. The social perspective of organizational culture is crucial for efficiency and agility. A lack of implementation of pro-social orientation in many cases is a barrier to the development of modernly managed companies. In turn, the lack of progress resulting in technological debt related to the lack of implementation of modern solutions reduces the level of competitiveness of enterprises and does not bring them to the level of business platformization, i.e., the transfer of businesses into the space of complex systems cooperating with other complex systems based on technology fusions. The interdependence of different technologies provided by different suppliers is a condition for creating solutions that are interoperable with commonly used technical solutions. Thus, it is now possible to build efficient and effective business models in the digital economy that function in specific business ecosystems.

4.3 A BUSINESS ECOSYSTEM IN THE DIGITAL REALITY

The business ecosystem is a relatively new concept that has been widespread in recent years. The social impact of modern enterprises, increasingly measured by their carbon footprint, requires managers to include it in the monitoring system. The business ecosystem is based on the assumption that the company operates in a specific external environment that has a greater or lesser impact on the environment and in particular, as has been strongly emphasized in recent years, on climate change. Climate change is manifested by an increasing proportion of anomalies in areas such as weather, violent storms (including thunderstorms), droughts, flooding, changing sea fronts, melting glaciers, etc. In addition to global conditions at the micro level, the activities of companies affect pollution, including emissions to soil, land, air, waste consumption, but also water management, including meteorological, agricultural, hydrological and hydrogeological droughts. It is also crucial to take into account biodiversity and the ecosystems discussed. Innovation should take place in the areas of resource use and the circular economy. However, the greatest challenge arises when we think about how a company's activity can have a positive impact on climate change. The topic is difficult and requires a great deal of analysis and shaping of new cognitive spaces in this area. This ecosystem requires good recognition in ontological and epistemological terms. In the common understanding of the business ecosystem, on the one hand, it can be treated in terms of sustainable development, social impact and such management where the pro-ecological orientation is of key and strategic importance; but on the other hand, it can be treated universally as simply the environment in which the company operates today because the business environment is influenced by various actors who are focused on pro-ecological policy, but also those whose goal is to maximize the market value of the company, to maximize profits at the expense of society. The search for optimal solutions is needed in which it is not only about creating economic profit, but also about creating social profit to ensure that key actors around the functioning of the company will not hinder its development and growth for reasons that do not meet the ecological criteria or the implementation of anticipatory and prognostic strategies which assess the ability of companies to adapt to climate change. Certainly, the change of consumer needs stimulates the creation and functioning of business in the conditions of the business ecosystem, which is a premise of pro-ecological behavior of companies and the contrast between two different business models of a modern company. One is an approach to business where environmental responsibility is recognized as one of the priorities for action on the market, and the other is more focused on reducing the unit cost in mass production using all possible ways, even if it leads to environmental degradation and adversely affects climate change that leads to weather anomalies and other negative environment conditions felt by humans, animals and plants. In order to understand this subtle difference between the two approaches well, it should be explained that here the two trends clash in terms of technological, organizational and social progress. A positive factor stimulating this type of solution, especially in the area of functioning in the business ecosystem, is the promotion of optimal solutions and overtaking the competition by implementing innovative solutions in this area. The business ecosystem is therefore based on the assumptions of, among others, interfaces between the environment where the company operates, the technologies used and the company's strategy embedded in the configuration of modern business models such as inclusive, progressive, and sustainable business models, etc. The business ecosystem is therefore a choice that managers make in order to follow modern trends, increase efficiency and effectiveness, reduce unit production costs and achieve financial success measured by economic and social profit, while obtaining a positive response of stakeholders as key actors conducive to or hindering the growth of specific companies. A slightly different issue in terms of understanding the concept of the business ecosystem is the virtual digitalization-based environment. In this sense, this phenomenon creates the concept of ecosystems. Then, it is important to create a focus on success supported by digital solutions in areas such as customer focus, integration of multiple value chain ecosystems, Big Data analysis, real-time technologies and other technical innovations. Digitalization supports numerous aspects of life and business. Business must change, otherwise it will omit key points on the timeline, the consequences of which may be negative in terms of the strategic development of companies and result in the loss of future prospects.

4.4 A TECHNICAL ECOSYSTEM AS A PLATFORM TO BUILD A NETWORK OF CONNECTIONS

Solutions shaping digital business models in the era of digital transformation supported by the use of artificial intelligence solutions, the Internet of Things, and the Industry 4.0 concept stimulate innovative and creative solutions by creating a technological business ecosystem. This business ecosystem is quite different from the classic business space, which was based on the traditional linear value chain. The metaphor of the technological ecosystem comes from the field of biology and has been transferred to software development because it reflects the evolutionary nature of software well. A large number of authors use the definition of a natural ecosystem to support their own systems of defining a technological ecosystem (García-Peñalvo 2018, p. 1). Such a technological ecosystem is based on many solutions on digital platforms. Their characteristic feature is that they operate in a network environment, and their existence depends on the number of interfaces, as well as their effective use in the performance of core functions that create value for the customer or support the overall system. Building a network of relationships in the technological environment is based on innovative algorithms, IT solutions, and all this is integrated into the logic of doing business, and more specifically, the way in which the business model delivers value to customers and other business actors. The technological ecosystem based on platforms creates opportunities for the unification of solutions, which means that it accelerates the implementation of design solutions. Rapid technological development gives new opportunities to assess ecosystem services, which can help overcome current knowledge gaps and limitations in data availability. This can lead to better integration of different technologies in the future, e.g., Earth observation, data science and online platforms, as well as stronger transdisciplinary cooperation to improve knowledge on ecosystem services and broaden the perspective on technological development in other related fields of research (Schirpke et al. 2023, p. 1). There are practical implications for members of innovation ecosystems, such as technology and knowledge-intensive entrepreneurs, social entrepreneurs and technology and knowledge transfer officers (Gerli et al. 2021, p. 1). However, attention is paid to the lack of clarity as to where the ecosystem perspective adds value and where it does not, which hinders the usefulness of these important ideas (Adner 2017, p. 39). In terms of the market, the platform should primarily be seen as digital infrastructure that connects producers, consumers, service providers, advertisers and other groups in an interactive ecosystem and enables the exchange of goods, services or information. The strategic goal of such a platform is to dominate all social and hybrid relationships between humans and "non-human actors" within the micropayment space, i.e., profiting from the interaction between people, organizations and things (Kreft 2021, p. 25). The ability to use technology, including the use of complementary technologies of business partners, creates a subjective relationship between integrated solutions of various suppliers in order to deliver optimal value to the customer. The network of relationships is the foundation of the effectiveness and efficiency of network business models. Network business models are solutions in which the condition for creating the expected function is the coexistence of at least two other business models with the consistency of technological, organizational and business aspects. The network and its attributes determine the competitiveness of the network business model. In turn, the technological aspect used in the environment of complex systems is focused on the implementation of the purpose function understood in business and operational terms based on processes and projects. An ecosystem in such an approach is understood as an environment for doing business using the social relationships of the main actors of networks supported by technological platforms that enable cooperation. The ecosystem in terms of technology should be characterized by scalability, which is the basic condition for achieving economic effects while maintaining the continuity of the entire system. The system must be constantly effective even when many new users are added to it. It loses this ability by ceasing to be scalable when it is overloaded with too many elements that make up the system. Designers and technologists of such a system must constantly take care not to lose the scalability attribute, i.e., the ability of the system to perform core functions when adding subsequent users to the system. Any dysfunction of a socio-technical system based on scalability is a fault of the entire system visible from the perspective of not only the system administrator, but sometimes its recipients. Therefore, two layers of systems that must be coherent can be distinguished, namely the layer of users that will always be finite and dependent on the technological capacity of the system and the technical layer depending on the solution used. Solutions are effective when there are no disturbances and anomalies between the network layer of relationships and the operational maintenance of the system. Any such event may risk the system losing the ability to perform the expected tasks.

4.5 ARTIFICIAL INTELLIGENCE AND DIGITAL BUSINESS—THE SOCIO-TECHNICAL APPROACH

Artificial intelligence is a very capacious concept and is considered from an infinite number of perspectives. It seems reasonable to say that artificial intelligence now and in the future will change basically all aspects of life, including business. It is used in many areas of business activity, in particular to handle processes that can be easily standardized and based on specific algorithms. The multitude of applications of intelligence as mentioned above is limitless and stems solely from the imagination of designers of business models and processes. Artificial intelligence in the case of multi-sided platforms is a condition for their effective functioning. In business, artificial intelligence analyzes Big Data sets and can thus be used for evidence-based management. This approach unleashes new space for managerial analysis, focused on capturing this information from among many variables, which through specific algorithms will allow better, in-depth support for managers in building long-term strategies and in current operational management. Functions such as speech recognition, image processing, music creation, art creation, dataset analysis and the like are used to create new areas of business and support current ones. Feeding such business models with artificial intelligence increases their capabilities and creates new functionalities. They are fundamentally important for the development and support of socio-technical systems. They set new opportunities and, consequently, standards for the implementation of work processes, which is crucial in optimizing supply chains. Through dedicated algorithms, it is possible to optimally manage production, warehouse and logistics processes. The efficiency, effectiveness, and agility of an organization depend on the interdependent relationship between social and technical factors. Artificial intelligence in these areas can be an extremely important support for achieving these parameters. Simple coupling and feedback loops within socio-technical systems, derived from the assumptions of cybernetics, introduce the need to develop new algorithms for the use of artificial intelligence. The development of information, databases and other systems included in the factors that stimulate digital transparency is not only a technical issue, but also a matter of business organization, which concerns in particular the process of change (Shona and Warren 2010, pp. 189–197). Artificial intelligence is an example of a tool where the IT issue is less important than the functionalities that this tool is supposed to generate in a specific situation. Artificial intelligence, with its specificity, development of its performance and functions, increasingly fits into the processes of support for achieving the abovementioned business indicators, namely the efficiency, effectiveness, and agility of the organization. AI is therefore not only crucial to gaining a competitive advantage over companies that are still some time away from developing this technology, but also to transforming many areas of citizens' lives. Socio-technical systems should be smart. The design of intelligent socio-technical systems requires the careful study of relevant social and organizational concepts. Research assumptions are based on generalizing certain observations of phenomena in order to build a theory that will underlie the construction of an artificial system. For the purposes of its testing, predictions resulting from the theory are created. The result of using the synthetic method is the construction of an artificial system such that the resulting animation, experiments or performance are then used to support or refute the theory (Jones et al. 2013, p. 5). The developed theory is used to test it in application conditions. This is how solutions stimulated by science are developed. To this end, artificial intelligence-based solutions are increasingly used. Artificial intelligence can also be evaluated through the prism of values, which Ibo van de Poel draws attention to. When implementing AI solutions, consideration should be given to issues such as respect for autonomy, harmlessness, integrity, transparency, explainability and accountability. In this approach, ethical influence agents are robots and computer systems that ethically influence their environment. In addition, robots and programs should be programmed (by humans) to behave according to certain values (van de Poel 2020, pp. 385–409).

These values give meaning to the systems designed and provide logic relevant to human reason. Therefore, an increasing number of AI technologies are being used in companies powered by artificial intelligence. These include, but are not limited to, supervised machine learning implemented through the creation of predictive models, neural networks that use hidden layers for prediction, deep learning that uses multiple hidden layers to create predictive models, image recognition that allows for the recognition of images in tagged data sets, deep learning—natural language processing that helps understand speech and text production, and robotic process automation (RPA) that combines workflows and data access, etc. (Davenport and Mittal 2023, pp. 28–29). Modern enterprises based on socio-technical systems which use artificial intelligence are turning into digital vehicles inscribed in the ecosystem of current business. They drive economies, change the rules of the game in sectors and play the role of game changers in many situations. This is the only direction that seems reasonable to achieve success by creating new value for customers and other stakeholder groups.

The position of artificial intelligence in the structure of the business model is presented in Figure 4.1.

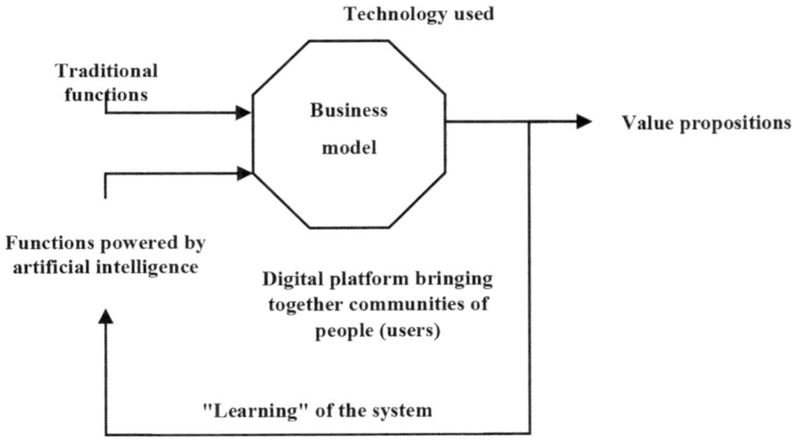

Figure 4.1 Location of artificial intelligence in the structure of the digital business model. *Source:* Own study.

The business model based on the digital platform can be powered by artificial intelligence algorithms that support its traditional functions implemented by means of the technologies used. The levels of value delivery matching are enhanced through dynamic feedback loops based on the assumptions of cybernetics theory.

4.6 SAFETY AND CYBERSECURITY AS A FACTOR IN SOCIO-TECHNICAL SYSTEM STABILITY

In companies where socio-technical systems stimulate the configuration of business models, if they are high reliability organizations, the issue of technical safety and cybersecurity is crucial. Taking into account the correct understanding of these concepts in the logic of running a business forces managers to build awareness and knowledge in this area. Some sectors of the economy, due to their specificity, must base their activities on fulfilling stringent safety requirements. Technical safety and cybersecurity management mechanisms should be built in their business models. Such a socio-technical system will require these issues to be addressed in the rules of conduct described in the relevant procedures and instructions. Technical safety is directly related to operational processes, where humans, as the operators of technical systems, are part of these systems and affect safety in the same way as technical elements. Hazards resulting in adverse events can come directly from human actions (errors, negligence, lack of knowledge, poor interpretation of the situation, etc.) and equipment defects. There are also different areas, traditions and concepts within the general field of human factors and ergonomics (HFE) that deal with the design, planning and management of human interactions within complex work systems. These include, but are not limited to, macroergonomy, cognitive systems engineering, sociotechnical systems theory, anthropotechnology, work psychodynamics, human system integration, organizational culture and safety climate, and resilience engineering (Carayona et al. 2015, p. 548). System safety is affected by the general characteristics of socio-technical systems. The socio-technical system includes normative laws and principles, as well as other rules, defining legitimate or acceptable uses of technology, relevant or legitimate owners and operators, places and times of its use, ways of distributing the profits and burdens (and risks) arising from the use of technology (Burns and Machado 2009, p. 11). When we say that we focus on the materiality of technology, we refer to the ways in which its physical and/or digital materials are arranged in specific forms that will survive despite differences in place and time. Such a definition implies that the usefulness of the term "materiality" lies in the fact that it identifies those constituent features of the technology that are (theoretically) accessible to all users in the same way (Leonardi 2012, pp. 25–48).

In the context of the specificity of such organizations, the role of cooperation between man and machine in creating technical systems is strongly visible because it determines the possibilities of ensuring the expected level of safety. Safety in this context is built both in the processes carried out as part of the activities of companies and in products, devices, systems, computer devices and software. In turn, a person acting as an operator should meet all the criteria of proper service both in terms of technical knowledge and the required level of concentration based on organizational factors and meeting psychophysical characteristics. The human and organizational factor (the organization of people's work in the system) should ensure the proper planning of processes in such a way that machinery and other equipment do not pose risks during their operation. Social engineering systems are becoming more complex and increasingly automated. However, human error is still widely seen today as playing a key role in most system failures (Hodgson et al. 2013, p. 608). There should be a clear separation between the understanding of technical safety and cybersecurity. Cybersecurity refers to external factors which the system is exposed to, for example, malicious entities or persons who wish to harm the proper functioning of the system through external interference with its functions. IT protection and physical protection applied by companies should protect them from interference by third parties who intend to damage the system or make it function incorrectly. Business models of companies that operate taking into account the assumptions of socio-technical systems should therefore implement a joint strategy for the implementation of protection against the effects of internal process errors (technical safety management and third-party attacks on the technical system—protection against cyber threats). For example, such systems based on socio-technical systems, where innovative business models are shaped, include the railway, sewage, energy and IT sectors. Within these sectors, business models should be designed in such a way that the relationship between man and devices is designed so that any threats arising in the course of this relationship have risk control measures in place to protect the organization from undesirable events. For each of these hazards, in the context of the probability of its occurrence, the risk must be estimated. Achieving an appropriate level of risk acceptability is a condition for allowing the process to be implemented. In the case of cybersecurity, all security systems, procedures, IT, hardware and physical security should be implemented and the degree of their effectiveness monitored. Wherever man plays a crucial role in the proper functioning of the technical system, the aspects of technical safety built in operational processes and protection against cyber threats should be taken into account. This is a condition for the recognition of a socio-technical system as efficient and effective in achieving the assumed objectives built into the designed functions of a specific action.

4.7 CONCLUSIONS

There are three key areas in technical systems, namely the technical system, the role of the operator in the technical system and the work environment. The relationship between man and machine must be designed in such a way that technology supports humans in making decisions, but does not remove responsibility from them. The final decisions are made by someone who, when supported by technology, is led by streams of technological activities in many situations. Then, thanks to the support of technology, he makes decisions while subject to pressure, opportunities and limitations. Eco-business systems in modern business depend on the efficiency and effectiveness of technology. They also depend on their life cycle, which has a significant impact on their service life, durability, reliability and safety. The analysis and diagnosis presented in the chapter showed that designing technical systems from the perspective of the functionality of business models is not easy and requires a number of conditions based on the theory of complexity.

REFERENCES

Adner, R. 2017. Ecosystem as structure: An actionable construct for strategy. Journal of Management. 43(1): 39.

Burns, T.R. and Machado N. 2009. Technology, complexity, and risk social systems analysis of risky socio-technical systems and the likelihood of accidents. Sociologia, Problemas e Práticas. 9(61): 11.

Carayona, P., Hancock, P., Leveson, N., Noyd, I., Sznelwar, L. and van Hootegem, G. 2015. Advancing a sociotechnical systems approach to workplace safety – developing the conceptual framework. Ergonomics. 58(4): 548

Davenport, T.H., Mittal, N. 2023. AII in on AI:. How smart companies win big with artificial intelligence. Deloitte Development LLC.

Gerli, F., Chiodo, V. and Bengo, I. 2021. Technology transfer for social entrepreneurship: Designing problem-oriented innovation ecosystems. Sustainability. 13: 1.

García-Peñalvo, F.J. 2018. Technological ecosystems for enhancing the interoperability and data flows. Information Technology Research. 11(1): vi–x.

Hodgson A., Siemieniuch C.E., Mae Hubbard E. 2013. Culture and the safety of complex automated sociotechnical systems. IEEE transactions on human-machine systems. 43(6): 608.

Jones, A.J.I., Artikis, A. and Pitt, J. 2013. The design of intelligent socio-technical systems. Artificial Intelligence Review. 5.

Kreft, J. 2021. Władza platform. Za fasadą Google, Facebooka i Spotify. Universitas, Kraków. 25.

Leitch, S. and Warren, M.J. 2010. ETHICS: The Past, Present and Future of Socio-Technical Systems Design. pp. 189–197. *In*: Tatnall, A. (ed.). History of Computing. Learning from the Past. 2010. IFIP Advances in Information and Communication Technology, vol 325. Springer, Berlin, Heidelberg. https://doi.org/10.1007/978-3-642-15199-6_19.

Leonardi, P.M. 2012. Materiality, sociomateriality, and socio-technical systems: What do these terms mean? How are they related? Do we need them? pp. 25–48. *In*: P.M. Leonardi, B.A. Nardi and J. Kallinikos (eds). Materiality and Organizing: Social Interaction in a Technological World. Oxford: Oxford University Press.

Schirpke, U., Ghermandi, A., Sinclair, M., Van Berkel, D., Fox, N., Vargas, L., et al. 2023. Emerging technologies for assessing ecosystem services: A synthesis of opportunities and challenges. Ecosystem Services. 63: 1.

van de Poel, I. 2020. Embedding values in artificial intelligence (AI) systems. Minds and Machines. 30: 385–409. https://doi.org/10.1007/s11023-020-09537-4.

Chapter **5**

Technical Assumptions of Digital Business Models based on the Assumptions of Socio-technical Systems

5.1 INTRODUCTION

As mentioned earlier, contemporary digital business models are based on the assumptions of socio-technical systems. Business platformization is conducive to the merging of the social and technical spheres. Only the joint implementation of social aspects with technical solutions makes it possible to achieve synergies between them. In this common sphere, business creators have found space to create a value proposition resulting from linking the needs of users with service providers via a digital platform. Special algorithms of socio-technical models and engineering methods enable the conceptualization and operationalization of digital business models. Current technological solutions make it possible to serve mass customers in a social system that, by interacting with technologies, exploits the potential of the technical system. In such reasoning, it is possible to create new forms of business based on technology. Schumpeter's theory of creative destruction (Schumpeter 1936) fits in this area in a special way. Business platformization is changing the game in almost all sectors of the economy. The concept of business models is included in the structure of the use of modern information technologies entering into interfaces with various systems

DOI: 10.1201/9781003456421-5

and subsystems based on technology fusions, exchange of knowledge and solutions based on the mechanisms of standardization and unification of data flow in order to achieve specific functions of business operations. The condition for the ability of an innovative solution to bring about changes in the sector based on the so-called *Game Changer* is the application of creative destruction to current solutions. These dynamically implemented business models reconfigure the structure of competitiveness in the sector, break the principles of accepted logic and set new standards and mechanisms of action in delivering value to customers. This is achieved by means of business virtualization which creates prospects for the development of digital innovations by creating a new digital business. Big Data sets and their adaptation for the flow of knowledge, information and communication help determine multidimensional possibilities for creating interdependent relationships between static data and dynamic users. The interdependence between these two areas creates opportunities for entrepreneurial action towards new business spaces.

5.2 SOCIO-TECHNICAL MODELS, ALGORITHMS AND ENGINEERING METHODS AS TOOLS FOR SHAPING DIGITAL SOLUTIONS IN THE DIGITAL ECONOMY

Digital twins, as a concept inscribed in an innovative approach to simulation, are extensively used in numerous development projects. The relevant literature defines at least 11 categories of business models with the support of digital twins (DT-BM): The Intelligent Product and Service System (PSS), the Circular Business Model (CBM), Intelligent Maintenance, Performance and Usage-Based Contracting, Freemium with Additions, Intelligent Manufacturing, Data-Based Digital Services, Subscription, Collaborative BM, the Metaverse Platform and Urban DT-BM (Adelsberger et al. 2024, pp. 409–414). Digital twins are based on the representation of reality in simulation conditions. They enable the testing of various solutions in an environment relevant to reality based on real data. The dynamic transfer of information from the field is conducive to learning, taking into account the variability of conducted processes. The concept of digital twins is used to optimize production processes and service delivery in the context of aspects such as machine control, predictive maintenance, and process improvement. Digital twins are used to build digital models based on real data. This function of digital twins allows for the transfer of simulation services to the real world and vice versa. Incorporating digital twins to create innovative solutions and improve processes enhances aspects such as safety, reliability, quality, efficiency, effectiveness, and the ability to change. Digital twins are based on algorithms that create virtual reality involved in the processes of the real world. Digital business algorithmization is the domain for designing digital business models. Reality descriptions transferred by algorithms to the virtual world allow for the testing of difficult situations in the context of various scenarios. Tasks are created for specific needs of specific sectors of the economy, the specificity of production or services provided. By means of simulation in relation to reality, the possibilities of improvements are determined. Digital twin models can be multidimensional and interdisciplinary across multiple scientific disciplines.

5.3 CYBERNETICS AS A TOOL FOR IDENTIFYING AND SHAPING CORE COMPONENTS AND THEIR ROLE IN CONTROLLING TECHNICAL SYSTEMS FOCUSED ON ACHIEVING THE EXPECTED FUNCTIONALITIES AND PERFORMANCE NECESSARY FOR SERVICING MASS CUSTOMERS

The theory of cybernetics should be considered in the context of the determinants of designing digital business models taking into account the assumptions of socio-technical systems. The examination of cyber solutions in the field of network theory is particularly important. In this respect, building complex systems connected by interfaces and controlled centrally, or in some cases decentralized, should be supported by knowledge and experience in the field of cybernetics. Norbert Wiener defined cybernetics as the science of control and communication and applied this assumption to machines and living organisms (Wiener 1948). The theory of cybernetics was also developed by W. Ross Ashby (Ashby 1956) and Claude Shannon (Shannon 1964). The premise of cybernetics is to transform information into control systems. Cybernetics is based on the assumptions of technical logic, which derives from mathematical logic. It is based on the theory of information, data transmission and the transformation of signals into specific functions that are useful to the recipient of this information. In particular, it is based on the theory of algorithms, which contains the principles of logical data processing in a specific cycle of planned tasks. Based on algorithms, IT solutions are created that transform the designer's intentions into specific functions of organizational and technical systems. In this way, self-organizing systems are created, often supported by artificial intelligence.

Cybernetics is based on simple coupling and feedback loops. In this way, the principles of complex systems are developed, from and on the basis of which digital business models are designed and built. The basic task of cybernetics in this aspect is also to shape effective and efficient, market-oriented, innovative digital platforms based on the assumptions of socio-technical systems. The control object is then a business model understood as a complex system of interconnected parts focused on the achievement of a specific goal. A complex system of interconnected parts that maintain strong cohesion among themselves, embedded in a specific management intention of the creators, meets the expectations of the market and technical progress. In this way, innovative solutions of modern digital business are created. The business model as a static element, thanks to the assumptions of the theory of cybernetics, becomes a dynamic system controlled by and triggering the interaction of many of its elements at the same time. Cybernetics in such a system should be embedded in the assumptions of socio-technical systems, taking into account modern knowledge of digital platforms and tools supporting information exchange, including real-time information exchange.

The theoretical model of system control in a cybernetic approach, based on which the principles of the functioning of digital platforms can be examined, is presented in Figure 5.1.

The presented diagram indicates the relationships between various functions of the digital platform control system described from the cybernetic perspective. In several core places, various algorithms are included to control the socio-technical system in such a way that all core functionalities of the platform are provided to users (the community) based on the logic of value creation.

5.4 USING A SERVITIZATION CONCEPT FOR DIGITAL TRANSFORMATION AS A DYNAMIC PROCESS OF SOCIAL AND ECONOMIC CHANGES OF BUSINESS MODELS BASED ON INNOVATIVE SOLUTIONS

The blurring of the boundaries between products and services is the result of focusing on business models that provide a platform for the implementation of product and service business systems. Business ecosystems are changing, exposing a strong impact on the improvement of customer service processes. Product and service systems are a combination of products and services that are systematized to provide the desired usability or functions that satisfy the needs of customers. The role of implementing the concept of a customer-centric organization

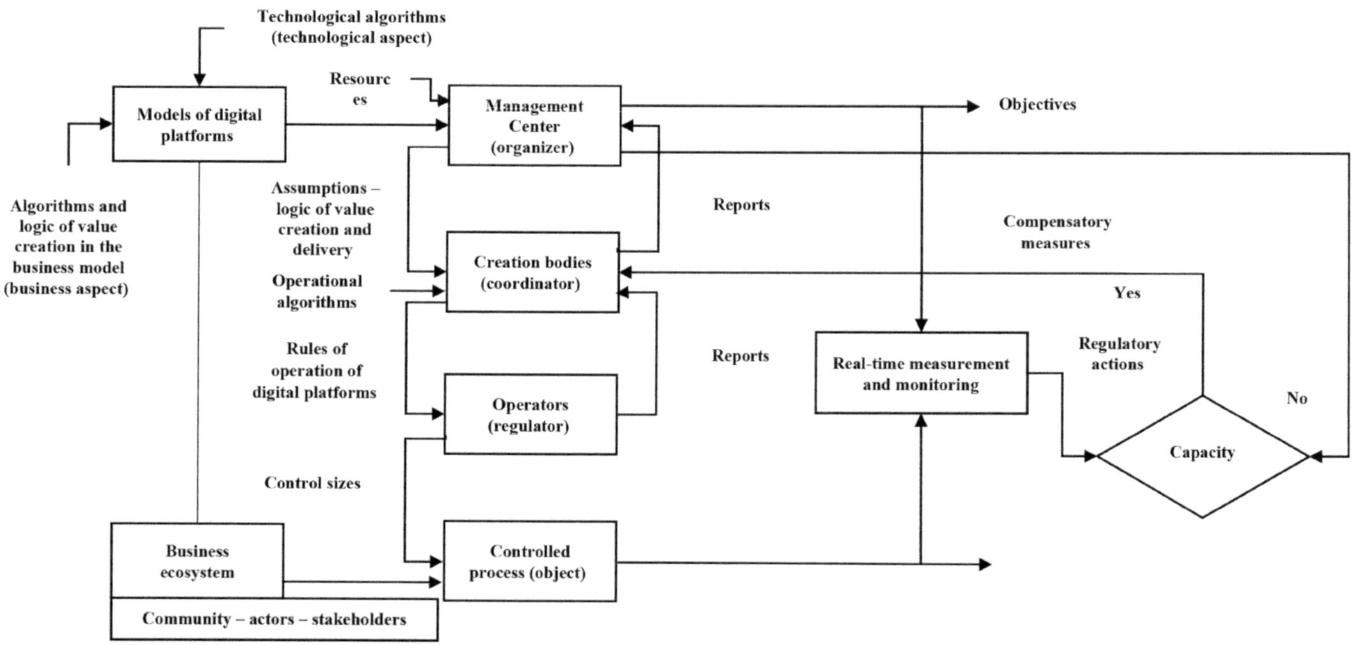

Figure 5.1 Program for controlling the digital platform as a socio-technical system from the cybernetic perspective.
Source: Own study based on J. Gościński (1973), Cybernetyczne podstawy informatyki [Cybernetic fundamentals of computer science], OBRI, Warszawa.

is growing, whereby all activities in the organization are focused on delivering the best value for the customer. Product–service systems (PSS) in this context propose an individual approach to a specific solution that provides the opportunity to fulfill even the most specific expectations and needs of customers and increase the competitiveness of companies (Long et al. 2016, pp. 5337–5360). The concept of servitization emerges from the definition of product–service systems (PPS), which stimulates the emergence of innovative business models. In this formula, their entire configuration, even if it belongs to a production sector, is focused on serving the customer in such a way that all the activities in this circular value chain are subordinated to the expectations and requirements of customers. In 1988, the term was defined in scientific literature by Vandermerwe and Rada (Vandermerwe and Rada 1988, pp. 314–324). Servitization is a phenomenon observed especially in recent years, which results from the dynamically growing role of services in the global economy, with a focus on the use of services within households, as well as extending supply chains by offering additional value integrating the supply of products with after-sales service. Factors that influence the development of the service sector, globalization, the mobility of factors of production, the decrease in transaction costs and the diffusion of technological progress play an important role. The development of artificial intelligence, the Internet of Things and modern data transmission systems is conducive to these processes. These determinants create the internationalization and fragmentation of production and increase the dependencies between economic sectors. Servitization is defined as the transition from a "clean" product to a "clean" service (Smith et al. 2014, pp. 242–269). A service is any action that one party can offer to another and that is intangible. Traditionally, this concept determines the transformation of supply chains into service chains, while in the case of innovative business models, servitization is already embedded in the fundamental idea of shaping these business models. Servitization is the innovation of an organization's capability and processes to change from a focus on selling products to selling integrated products and services that provide value in use (Baines et al. 2009, pp. 547–567). Features of services that are crucial in the context of the technological approach include intangibility, inseparability (services produced and consumed at the same time), diversity, and the inability to store. In particular, Vargo's concept of service—dominant logic (Vargo and Lusch 2006, pp. 43–56) stimulates new formulas for defining and understanding the service orientation of companies. In this concept, business is focused in such a way that the key factor in the effectiveness and efficiency of the business models is perfectly provided services. Thus, the heterogeneity of the context affects the way resources are used for service. In this approach, the designed systems are configurational, combining exogenous and endogenous factors into one common system, as is the case with the combination of social and technical systems.

The business model is the embodiment of a specific configuration due to the arrangement of its components to ensure the specific formula of creation and delivery of value to customers. The servitization process is the use of the company's parallel product and service-oriented business models (Palo et al. 2018, pp. 486–496). The process of servitization thus understood can be a way to transform the business models of companies from product-oriented to service-oriented.

5.5 AUTOMATION AND ROBOTICS IN THE PROCESSES OF SHAPING VIRTUAL DIGITAL BUSINESS MODELS

Automation and robotics, especially in the area of manufacturing companies, significantly improve operational processes. Production processes, but also services, are changing, taking into account the assumptions of business servitization. In both production processes and services, digital solutions are shaped that improve the processes of direct production of products or services, but more broadly refer to the entire business model of the company. Automation in production systems should be built taking into account the assumptions for socio-technical systems. Unfortunately, there is often an 'either/or' attitude towards people and technology, which tends to focus on optimizing separate human and technical components, and interface design and sharing is secondary (Behymer and Flach 2016). Digital transformation determines the possibilities of creating virtual digital business models supported by automation and robotization processes. Delegating tasks to a non-human agent in a social engineering system can expand human capabilities and performance. However, effective performance depends on the successful relationship between human operators and automation (Roberts et al. 2021). Human factors have the strongest impact on the development of digital business models on the web. In this way, they create sociotechnical systems of value creation (Trabert et al. 2022, pp. 471–481). Technical and organizational efficiency then depends on algorithms and the connections within them of machines, devices and operators managing these processes. Automation and robotization make the functionality of business models dependent on their capabilities. The Internet of Things, data transmission, cybersecurity, technical security, artificial intelligence and other innovation-driven solutions are essential to deploying solutions that generate progress and improve business efficiency. Technologies embedded in business models support the business orientation of enterprises with the technical conditions for building a competitive advantage. The virtualization of digital business models is based on the fusion of solutions of various enterprises that cooperate to achieve a business goal and create new solutions using several parties to the contract. Technology fusions are currently the main trend based on which virtual digital business models are shaped. Automation and robotization are a prerequisite for these solutions to be useful in multiplying and scaling the business. The adaptation of new solutions to the conditions of shaped innovations based on the solutions of various market players significantly influences the production of groundbreaking business models. The configuration of such business models includes those features of functionality based on the traditional relationship with the customer and those features of functionality that are supported by modern technologies. The balance between traditional business concepts and the ability to transfer innovation is a prerequisite for the sustainable development of technology-oriented companies.

5.6 DIGITAL PLATFORMS BASED ON ALGORITHMS AND THE BIG DATA ENVIRONMENT IN LIGHT OF TECHNICAL MODELS BASED ON INNOVATION AND TRADITIONAL APPROACHES—CONFRONTATION

Digital platforms create innovative digital business models that meet the assumptions of socio-technical systems. In addition to the integration of technical and human factors related to statics, they are dynamized by algorithms that perform their assumed functions. Algorithms are called socio-technological compounds. There has been a conceptualization of algorithms for digital platforms. Algorithmic technologies are part of broader social realities and programs of socio-technical change; therefore, their design and development should be based on the interests and rights of users in the social, political and cultural environment (Shin 2019). A simple definition of algorithms refers to their understanding as procedures for converting input data into output data for a specific scheduled target function (Dahlman et al. 2021, pp. 1–15). Algorithms for their applications operate particularly on the basis of Big Data sets characterized by processing data very quickly and on an unprecedented scale. Big Data decision-making systems are 'worked' by large teams of people. The algorithm, from this broader perspective of business models, is no longer just an IT tool, but the way in which certain functions are performed at the interface between business and technology. Business algorithmization is a new space for creating new solutions, which is often called algocracy. Algorithms are interdependent and function in a network connection. Together they form conglomerates, which are complex interdependent systems operating in a data processing environment. This creates new formulas for delivering value to customers based on algorithmic processes. At the same time, a specific community is built and integrated with communication and transaction support tools. Through digital platforms, whose functions are based on algorithms, these transactions and information flows are initiated and executed. Algorithms control these processes and increase the intelligence of socio-technical systems (Jones et al. 2013). This intelligence is based on the functions and capabilities of the algorithms used together with their embedding in the process of creating value for users/customers. Digital platforms are dynamic organizations. Changes in the real state thereof are subject to modification depending on the needs of users and the degree of saturation of requests to perform certain operations in the system. In addition, all components of the system should be inter-integrated, especially in the context of storage and migration conditions. It should also be noted that the systems that create digital business models are fragmented. The data and information are located in different places and are used in accordance with the requests for their use. As part of technical systems, data, databases and information are integrated and created in such a way that a coherent technical system is created. The same applies to the social system, where user groups should be segmented into groups that are reasonably homogeneous in order to better manage them and 'use' them in the process of receiving value from them or often co-creating value. Then, the abovementioned integration should ensure the joint implementation of the assumptions of socio-technical systems on the basis of the functionality of digital platforms supported by appropriate algorithms. Algorithms are therefore used to develop conceptualization and, consequently, to operationalize the activities of digital business models. Mass databases are categorized for use and application to specific functions embedded in the working logic of algorithms. Their logic also applies to the functioning of social media specific solutions suitable to the digital economy. Algorithm intermediation in social relations appears to be a necessary condition for optimizing the effects of human activities enabling the achievement of business or social goals (Kreft 2019, p. 179). Therefore, algorithms support people and indicate potential solutions which might be suitable to a potential user. Algorithms are a key vector for digital platforms operating on the basis of socio-technical systems.

5.7 CONCLUSIONS

The content presented in the chapter based on technological assumptions in the process of understanding the functioning of digital business models has proved that, on the one hand, new technologies, computerization and networking are important for the development of this issue, but on the other hand, it should be based on scientific achievements, for example in such an area as cybernetics. Simple couplings and feedback are built into the process of designing business solutions using agile project management methodologies. They also concern technical issues in terms of transmission of the data stream, as well as the reception thereof and determination of the status and technical condition of the devices which are reconfigured. The concept of servitization, automation and robotization with intelligent algorithms creates the conditions for the implementation of innovative functions accelerating customer service processes, information transmission and business decision making. These contents formed the basis for the analysis of literature and practical experiences contained in this chapter of the monograph.

REFERENCES

Adelsberger, R.T., Antons, O. and Arlinghaus, J. 2024. Digital twins and their implications for business models: Overview and potentials. IFAC PapersOnLine. 58–19: 409–414.

Ashby, W.R. 1956. An introduction to cybernetics. London: Chapman and Hall.

Baines, T.S., Lightfoot, H.W., Benedettini O. and Kay J.M. 2009. The servitization of manufacturing: A review of literature and reflection on future challenges. Journal of Manufacturing Technology Management. 20(5): 547–567.

Behymer, K.J. and Flach J.M. 2016. From autonomous systems to sociotechnical systems: designing effective collaborations. Journal of Design, Economics, and Innovation. 2(2).

Dahlman, S., Gulbrandsen, I.T. and Just, S.N. 2021. Algorithms as organizational figuration: The sociotechnical arrangements of a fintech start-up. Big Data and Society. 1–15.

Gościński, J. 1973. Cybernetyczne Podstawy Informatyki [Cybernetic Fundamentals of Computer Science]. OBRI, Warszawa.

Jones, A.J.I., Artikis, A. and Pitt, J. 2013. The design of intelligent socio-technical systems. Artificial Intelligence Review.

Kreft, J. 2019. Power of Algorithms, at the Sources of the Power of Google and Facebook. Jagiellonian University Publishing House, Kraków. 179.

Long, H.J., Wang L.Y., Zhao S.X. and Jiang Z.B. 2016. An approach to rule extraction for product service system configuration that considers customer perception. International Journal of Production Research. 54(18): 5337–5360.

Palo, T., Åkesson, M. and Löfberg, N. 2018. Servitization as business model contestation: A practice approach. Journal of Business Research. 104: 486–496.

Roberts, A.P.J., Stanton, N.A., Pope, K.A. and Fay, D. 2021. To utilize automation or not to utilize automation, that is the question: An evaluation of how drills and procedures impact optronics mast usage from a sociotechnical systems perspective. Applied Ergonomics. 97: 103543.

Schumpeter, J.A. 1936. The Theory of Economic Development: An Inquiry into Profits, Capital, Credit, Interest, and the Business Cycle. Cambridge: Harvard University Press.

Shannon, C. 1964. The Mathematical Theory of Communication. Urbana: University of Illinois Press.

Shin, Don D.H. 2019. Socio-technical design of algorithms: Fairness, Accountability, and Transparency. 30th European Conference of the International Telecommunications Society (ITS): "Towards a Connected and Automated Society", Helsinki, Finland, 16th–19th June 2019. International Telecommunications Society (ITS). Calgary.

Smith, L., Maull, R. and Ng, I.C.L. 2014. Servitization and operations management: A service dominant logic approach. International Journal of Operations and Production Management. 34(2): 242–269.

Trabert, T., Beiner, S., Lehmann, C. and Kinkel, S. 2022. Digital value creation in sociotechnical systems. Procedia Computer Science. 200: 471–481.

Vandermerwe, S. and Rada, J. 1988. Servitization of business: Adding value by adding services. European Management Journal. 6(4): 314–324.

Vargo, S.L. and Lusch, R. 2006. Service-dominant logic: What it is, What it is not, What it might be. pp. 43–56. *In*: Lusch, R. and Vargo, S.L. (eds). The Service-Dominant Logic of Marketing, Dialog, Debate, and Directions, M.E. Sharpe, Armonk.

Wiener, N. 1948. Cybernetics or Control and Communication in the Animal and the Machine. Cambridge, MA: MIT. Press.

Chapter 6

Business Models Chosen as a Result of Using Socio-technical Systems

6.1 INTRODUCTION

Understanding the place and role of socio-technical systems in the construction of digital business models requires an understanding of a number of crucial theories that affect a full and comprehensive explanation of the phenomena of the functioning of the digital economy. The chapter contains an explanation of the assumptions of the theory of complex systems, the modularity of digital business, the role of knowledge in the codification and categorization of learning systems and indications of what new innovative business models are currently being implemented in business. Regenerative, sustainable, circular, progressive and inclusive business models are setting new trends in the search for attractive solutions in the field of customer service. They provide an opportunity for innovators to create profitable business ventures. Understanding business, technological and social relationships is the key to a contemporary approach to the entrepreneurial behavior of business creators. These contents are included in this chapter of the monograph.

DOI: 10.1201/9781003456421-6

6.2 COMPLEX SYSTEMS AND MODULARITY OF SYSTEMS—COMPLEX PRODUCTS AND A COMPLEX ORGANIZATION

Modern socio-technical systems are complex. They fit into the concepts of complex technical systems supported by relationships created as part of the construction of the business ecosystem. The "complexity" characterizes a feature of the system that arises from the interaction of the various components of the system and indicates that the behavior of the system cannot be reduced to the sum of the behaviors of its components. It should be treated as a whole (Daryani and Amini 2016, p. 359). The premise of this concept is that all organizations are networks, and social complexity shows that management is primarily about people and the relationships between them, not departments and markets (Appelo 2011, p. 30). Theories supporting complexity theory include dynamical systems theory (Barreira and Valls 2013), chaos theory (Kellert, 1994), network theory (Borgatti and Halgin 2011, pp. 1168–1181), the theory of evolution (Darwin 1859), game theory (von Neumann and Morgenstern, 1953), systems theory (von Bertalanffy 1929), cybernetics (Ross, 1957; Wiener, 1948, p. 212) and catastrophe theory (Thom, 1980), all of which aim to explain the behavior of deterministic systems (Grabowski and Wyciślak 2015, pp. 106). A crucial contribution made by complexity theory to management was to show that effective analysis and planning tools can be used to solve a variety of organizational problems. The basis for these innovations is network analysis (Nunes Amaral and Uzzi 2007, pp. 1034). Complex systems have many interrelated components that produce behaviors that cannot be understood by looking at individual components. Complex systems exhibit emergent features that result from the interaction of their components rather than from their independent features (Madanchian 2024, p. 2). These complex systems meet the definition of the System of Systems (SoS) concept (Ireland 2013, p. 248), which is based on the interdependent creation of a multidimensional and interdisciplinary system of network connections, where elements are mutually dependent and dynamically affect each other. Adaptive complex systems can both influence their environment and change their structure without external input. An important feature of complex systems such as self-organization is also emerging (Raaidi et al. 2018, p. 1). This proves that the system can function as if on its own by stimulating its own elements and connecting interfaces between them for the purpose of providing specific functions of the system. Complex systems such as transport, healthcare, and energy generation are inherently and inevitably dangerous (Cook

2020–2021). Business ecosystems based on digital platforms also meet the definition of complex systems and, at the same time, socio-technical systems that are aimed at building relationships with users and potential customers in order to build communities. Complex products and complex organizations in a complex external environment create complex systems that need to be managed. Each system, especially a complex one, is based on a modular approach, or at least this approach can be used to determine its configuration. System modularity is essential to creating interfaces with other systems. The complexity of the system in this situation is based on the compilation of numerous systems composed of a great number of modules with various functions. These functions may be related to basic tasks, but they may also be specialized, for example, safety functions, quality functions, cybersecurity functions or protection functions. Systems theory is used to determine the fundamental relationships between multiple elements which form a coherent uniform system with a specific purpose function. In addition to the core features resulting directly from the definition of the system, it is important to take into account the relationship between the factors that determine the operation of the system and safety functions. Quality functions that directly relate to the product attributes and application are also crucial. A system is a set of elements that interact with each other to achieve a specific goal. System elements include people, technologies, techniques, procedures, characteristics of objects and interfaces between all these elements. Machines forming part of a socio-technical system should be characterized by parameters such as reliability, availability, susceptibility to maintenance, safety, compliance with standards, quality, technical data and the like. In the context of interface relationships, the human–machine pair is crucial. In this system, by multiplying a certain number of people and a certain number of machines, a complex system is created where relationships take place dynamically to achieve the function of the goal that the system has to fulfill. Such systems may be focused on man having a greater share than machine or, conversely, machine playing a greater role than man. Therefore, any system in this sense will have to be identified and described. Appropriate behavioral algorithms should prioritize dynamizing and dynamized factors. Man and machine working together create a system of connections that has to perform specific tasks designed by the system designer. The logic of such a system should be included in the configuration of the business model taking into account the logic, methodology and conceptual frame, which should explain the assumptions of the complex socio-technical system used. For this purpose, models based on prediction are developed, i.e., predicting the future behavior of the system against the stimuli that affect it. Modularity is necessary to separate specific system components that can be described, have ontological features and can be selected and separated from the system, and a set of characteristics that can be managed can be built for it. System integrity will be achieved when all elements are focused on the implementation of core functions and compliance with the assumptions of designers. The state of a complex system will depend on the states of its individual components that perform various functions sometimes in series and sometimes in parallel, which has an impact on ensuring system safety. The set of states of the system in connection with the structure of the business model in terms of the dynamics of relations with the environment creates the possibilities of this system. Monitoring these states in the context of the functions of meta-technical business models combines technical, social, organizational and business aspects. Thus, sets of a complex system with a specific suitability are created. These states may have the nature of core useful functions of a product or service or may have the nature of complex systems based on satisfying criteria such as reliability, availability, readiness to use and others. When designing socio-technical systems from the perspective of complexity theory, the entire theoretical framework that supports such solutions with paradigms, constituting a set of rules shared by all specialists, should be used. It is crucial to model socio-technical systems with the core functions of the whole system as well as its components. In this way, system models are created that are relevant to achieving the planned functions. The social dimension in the context of the recognition of organizations as complex systems plays an important role in understanding and explaining all the relationships between technology and people. The sources of instability occur between them, that is, people and technology, but also between people when set against the background of technological solutions, which often lead either to conflicts that destroy a certain level of the status quo or introduce creative disorder leading to the development of new organizational solutions. At the same time, it creates opportunities for a new and creative approach to business activity. Predictability in modern times is very low; hence, the need to search for solutions based on the prediction of future events. Computer simulations make it possible to test, verify and validate the solutions developed. The control of complex systems supports functions of business models based on technological conditions. In this way, the business layer is complementary to the technological layer used in the defined social system. This approach determines the contemporary understanding of business in the conditions of the digital economy.

6.3 KNOWLEDGE CODIFICATION AND FUSION STRATEGY FOR COMPLEX MODULARITY SYSTEMS FOR BUSINESS PLATFORMS

Digital platforms are created by innovators who broadly rely on new ideas that break the rules of the market game. These are often "game changer" solutions, i.e., solutions that change the rules of functioning of sectors. Knowledge is a fundamental factor that stimulates the emergence of new business models. Complex systems based on modules, which contain solutions constituting the new know-how of their creators, can be used by knowledge creators as part of technology fusions. Knowledge is sometimes gained by making statements that serve to generalize a given data set. Information can be simple or complex in terms of its relation to the real world (Chiang et al. 2023, p. 18). There is a huge variety of current research on knowledge discovery in databases. Knowledge discovery in databases is feasible but requires the use of domain knowledge, efficient and incremental algorithms, interactive systems, and multi-level integration (Frawley et al. 1992, p. 58). Knowledge, therefore, is contained not so much in business models nor business platforms, but instead in modules that shape socio-technical systems. Socio-technical systems that underlie business platforms are based on a modular approach, where their complexity, consistency and ingenuity determine the success or failure of the proposed solution. Stimulating disruptive ideas in sectors are focused on specializations, based on which companies build competitive advantages and in many cases create new blue ocean business space. Technology, knowledge and people-to-people relationships built into the digital platform are able to create new space for delivering value to users, thus staying ahead of competitors' moves. The complexity of systems embedded in the business ecosystem based on socio-technical systems is conducive to the codification of knowledge, on the basis of which it is possible to expand product groups with new types, series and kinds. To this end, a variety of methods and technologies are used to facilitate and support knowledge sharing in learning communities (Zamiri and Esmaeili 2024, p. 1). The core competence, however, both in the technical and organizational sphere, is of paramount importance in the design process. The modern design of business models is based on the selection of resources from the sphere to digital solutions, software, and hardware available in technology, as well as the ability to capture the needs of large social groups. In this way, the migration of value to attractive business models whose complex socio-technical systems stimulate development is possible. This approach unleashes the potential of business creators, where managerial knowledge is essential in the sense of understanding markets, but technological knowledge serves to transform these needs into concrete solutions. Technology fusions in the case of technology-based business solutions are essential to managing development projects embedded in the modern business ecosystem. Modern technology fusions are therefore based on the exchange of knowledge unambiguously codified and belonging to a specific modular solution. The merger between Apple, which owns the iPhone, and manufacturers of automotive, usually luxury, brands as part of the CarPlay system is such a solution. The integration of digital solutions is therefore essential for the development of new products based on digital transformation, where specializations from other companies must be used. Therefore, a new era of creating cooperation strategies based on technology fusions inspired by the creators of new business models begins. The knowledge of some is a condition for the development of others. It is not possible in today's world to independently create solutions from the sphere of mechanics, electronics, electricity, electrical engineering, computer science and other areas of specialized knowledge. Therefore, modularity not only creates opportunities, but also forces the search for new cooperation partners and supports them in finding new solutions that can be implemented and create new market value as part of this fusion. Strategic alliances based on technology fusions are a necessary condition to build solutions that provide opportunities to create new and innovative solutions that also create the ability to capture value from the market. Through technology fusions, a business lever that pushes companies to significant market successes is noticeable. The potential in these fusions greatly exceeds the capabilities of an independent company. Even global giants in the digital economy are not able to create such solutions on their own that would be ahead of the competition. In this formula, meta-technological business models are created, containing all possible current functions that the customer expects. Technology fusions take place in the sphere of artificial intelligence, the use of Big Data sets, digital twins, or sectoral needs or specific conditions of technological progress. The integration of AI databases serves not only the effective application of artificial intelligence technologies and the development of database technologies, but also the next generation of computers that will support future intelligent information systems. They will make the work even more effective and productive. The integration of AI databases will generally contribute to science and technology infrastructures, businesses and humane computing applications (Austin et al. 2023, p. 161). Mobility, logistics processes, smart cities, renewable energy sources, data transmission and building technological interfaces are a necessary condition and a need for designing innovative solutions in the sphere of digital business models. Socio-technical systems in this situation are developed by these fusions, which the creators of new knowledge and liberators who break market rules stand behind. In science, however, after conducting a cycle of research and analysis, they are conducive to the emergence of new paradigms. The codification of knowledge must be process-oriented, where the processes of review, verification and validation of the designed solution must fulfill the requirements of quality safety. It is important to integrate knowledge, but also quality and safety functions built into modules, subsystems and systems. The digitalization of business, therefore, will change the current paradigms of understanding competitiveness towards cooption and cooperation. The technological needs of business development force this need for cooperation in the field of knowledge exchange, its codification, security and commercialization. New archetypes of products are being created and co-created by sharing and exchanging knowledge. New functionalities, the use of sensors, new forms of data stream, integration through interfaces and, in particular, the co-creation of knowledge determines the chance to create new ideas or withstand the competition. Therefore, products are created by technology fusions and business architecture is interdependent on other companies. Thus, analog solutions are transferred to solutions based on technology fusions. Knowledge between companies is transferred, which is a condition for mutual support of their businesses. According to Govindarayan and Venkatraman, the transformation of knowledge from analog to technology fusion occurs in four stages, namely building a fusion-based product architecture, organizing various processes supporting this action, using the acceleration effect, i.e., accelerating actions thanks to shared value, and finally monetizing the solution (Govinndarajan and Venkatraman 2024, pp. 82–83). Thus, the commercialization of fusions accelerates the process of monetization and distribution of value. In this way, the migration of value towards a company which uses this strategy is faster and more effective. Building modern approaches to highly competitive markets in modern business must be based on cooperation, trust and knowledge exchange. Business modularity, including the modularity of technological solutions, accelerates the progress and speed of creating new business models. Figure 6.1 shows the impact of technology fusions on shaping new digital business models or new technical functionalities useful for creating new business solutions.

New business solutions created in the form of creating new functions for innovative business models do not function in a vacuum. The external interactions of other systems stimulate the emergence of new space for technology fusions. Consensus in the conformity assessment of socio-technical systems may create synergy effects, which may result in the creation of a new business model, new functionality or usefulness

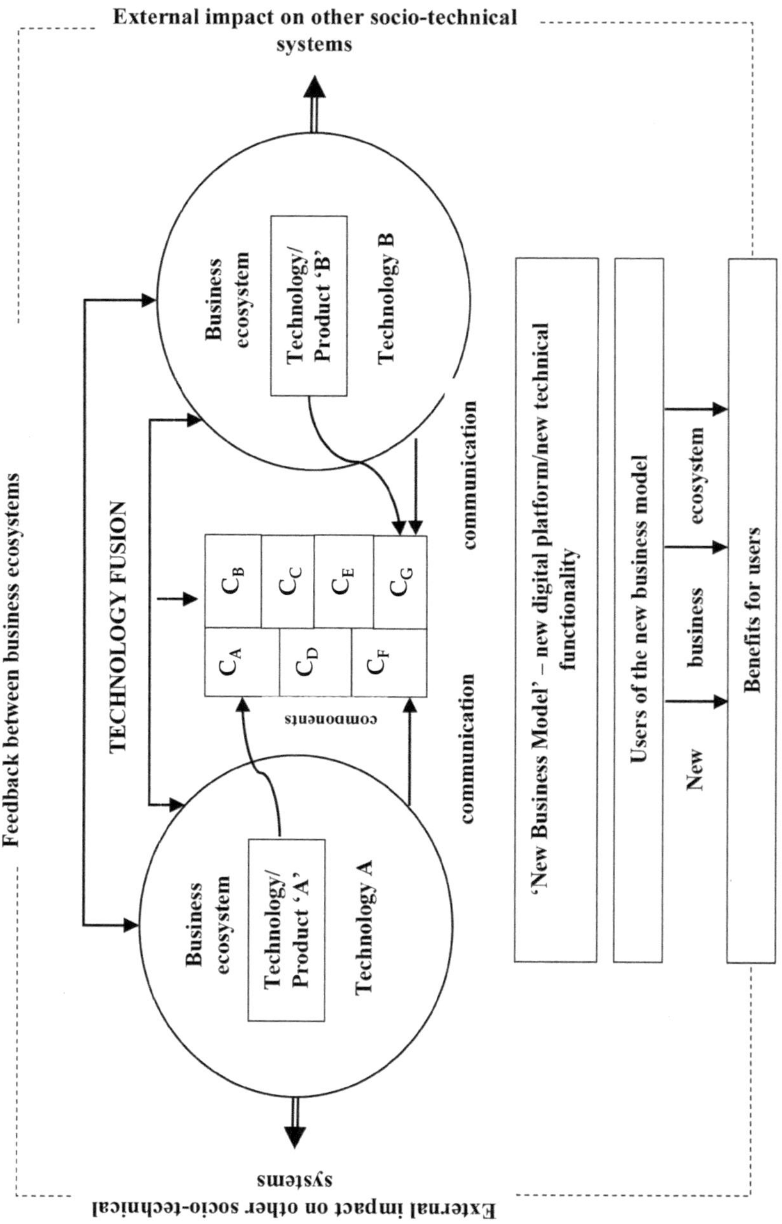

Figure 6.1 The impact of technology fusions on shaping new digital business models or new technical functionalities. *Source:* Own study

needed by other recipients. It can even be just a simple benefit, in demand in the technology market. Thus, a form of business monetization can be created for the benefit of many stakeholders.

6.4 REGENERATIVE BUSINESS MODELS AND SOCIO-TECHNICAL SYSTEMS

In the era of numerous social, political, economic, environmental and migration crises, companies need to build resilience. Resilience is a type of process that combines an organization's ability to adapt to dynamic change with its ability to set new strategies and reconfigure its business model after a crisis, disruption, or unexpected anomalies. Resilience, however, is a condition that helps prevent the negative effects of changes in external and internal environments to ensure the sustainability, ability to develop and growth of companies. Resilience allows for the maintenance of the core operational activities of companies in the face of the chaos that surrounds them. This ability means that companies that focus on implementing resilience strategies are better able to cope with difficulties and thus achieve a competitive advantage over competitors. Resilience is the first step to finding a way to function in very difficult environmental conditions and, consequently, to create value for shareholders and other stakeholders despite numerous limitations. Resilience can be created at the technical level, then cybersecurity or digital security—in the case of technical systems—can be referred to, but it can also be organizational in nature. Then, we believe that the organizational structures designed in the company and personnel authorized for individual positions are optimal and constitute the core foundation of the company's defense against threats. The next step that companies implement on the way to survival in conditions of chaos and crisis is the creation of regenerative business models. Then, companies search for effective and efficient ways to shift their business model from attributes based on resilience to a comprehensive regenerative configuration of the business model. The regeneration of business models is based in particular on the creation of a new formula for co-creating added value for customers, employees, the environment and the business ecosystem. Regeneration is based on the process of rebuilding and renewing elements of the business model configuration that have undergone complete destruction or loss of agency. It is the process of renewing the energy resources of employees, renewing financial strength and creating new elements of the investment attractiveness of the business model. In biological and cyber terms, regeneration can be compared to the return of the system—in this case, of the company to homeostasis, i.e., equilibrium. Homeostasis is the ability of a system to maintain certain quantities within ranges determined by the adaptability of the system, despite significant changes in external conditions. The phenomenon of homeostasis occurs thanks to the operation of self-regulation mechanisms involving the use of the feedback principle. In the case of external disturbances, an appropriate defense mechanism is activated that does not allow changes in internal states (Mynarski 1981, pp. 36–39). The resilience and regenerativeness of business models is of paramount importance to ensure the business continuity of companies exposed to negative impulses from the external and internal environment. If the internal environment is harnessed and all employees support the company's management in achieving survival goals, then a business model that is definitely customer-oriented, ideologically attractive and accepted by the environment will become regenerative. This regenerativeness will be characterized by the fact that, as the negative impulses from the external environment increase, the company will develop and face new, unknown challenges. The regenerative nature of business models is also particularly linked to the concept of sustainability, as it is based on balancing different business perspectives. The long-term survival of the business model must be based on the assumptions of converging concepts such as balance, equilibrium, homeostasis, etc. The idea in this regard is to ensure the satisfaction of all groups of stakeholders, but at the same time acknowledging that the company achieves success, which is a determinant of its further development. Socio-technical systems based on homeostasis apply the assumptions of the achievements of cybernetics such as sciences about the logic of control, which is currently regaining its lost youth. The achievements of cybernetics fit perfectly into the socio-technical assumptions, but also in the integration of strategic thinking with operational management supported by technical systems based on control.

6.5 SUSTAINABLE DIGITAL BUSINESS MODELS AND SOCIO-TECHNICAL SYSTEMS

The concept of sustainable business models has been widely studied in a number of areas for many years now. Not only are business issues included, but in particular engineering, climate, production and service issues, which are also assessed from the point of view of individual sectors of the economy. Digital business models, in turn, are based on digital platforms, in line with concepts such as the circular economy, the sharing economy or Big Data. Sustainability is currently based on the assumptions of Elkington's triple bottom line concept, which included three core pillars in its analyses, namely social, economic and environmental (Elkington 1997). Currently, this concept is developing under the name ESG, but in terms of reporting requirements of non-financial performance by companies. It is therefore a set of factors unrelated to purely economic development indicators that are included in the system of reporting results. Elkington paid particular attention to paradigm shifts related to modern management, defining seven old and seven new paradigms in seven core areas. New paradigms have been focused on areas such as competition, a soft approach to value creation, openness, a functional technology lifestyle, a symbiosis-based partnership, longer consideration time and the inclusion of corporate governance principles in decision-making processes. Table 6.1 shows the seven old and seven new paradigms developed by Elkington.

Table 6.1 Seven Old and Seven New Paradigms Underpinning the Concept of Sustainability

L.P.	Area	Seven Old Paradigms	Seven New Paradigms
1.	Markets	Compliance	Competition
2.	Values	Hard	Soft
3.	Transparency	Closed	Open
4.	Lifecycle technology	Product	Function
5.	Partnerships	Subversion	Symbiosis
6.	Time	Wider	Longer
7.	Corporate governance	Exclusive	Inclusive

Source: Elkington (1997).

The achievements of the sustainability concept are still developing and new concepts are emerging, such as ESG or regenerative business models. The idea of a socio-technical system, i.e., solutions in which the human factor interacts with technology and cooperates together to provide value to users, fits into the assumptions of the concept of sustainability. Technical factors have no value if they do not create a social ecosystem that supports partnership and contractual relations with the wider society. Due to the fact that mobile devices are common, the design of technical platforms based on the digital economy as a solution in the sphere of technical social systems allows for the design of such business models that fit into the concept of sustainability and build positive values for digital system users or digital solution providers. Business practice based on the theoretical assumptions of scientific achievements uses new areas of knowledge and searches for solutions that will help build trust between stakeholders, but in particular, accelerate the development and growth of companies for the benefit of not only shareholders but also other stakeholder groups. This acceleration would not be possible without conscious, wise and prudent observation of the internal and external environment of modern companies. Even when the logic of the system is embodied in technology, social and organizational factors must be taken into account. Then, the opportunity to create a sustainable, success-oriented company arises.

6.6 CIRCULAR BUSINESS MODELS AND SOCIO-TECHNICAL SYSTEMS

The circular economy is based on the assumption that the consumption of raw materials and the generation of waste is a fundamental aspect of the modern building of the efficiency of companies. Therefore, this assumption should be embedded in the business model of the company. It is the business model that should have a positive impact on climate change by reducing greenhouse gas emissions and the level of energy use, which stimulates the creation of new innovative production technologies or the provision of services. As regards circular business models, the value chain should not be linear, but circular, which means that the usefulness of the products produced, materials used or processed semi-finished products is maintained to the maximum limits. This sustainment extends the life cycle of individual components of the value chain, giving the opportunity to improve resource efficiency, including resources that would be considered waste in a traditional value chain. Conscious managers guided by social responsibility, but also economic rationality, are aware of the increased cost of raw materials or semi-finished products due to the shrinking of natural resources. Therefore, circularity stimulates the effective use of resources owned and unusable in traditional technology. Digital twins are elements crucial to developing the complex digital systems required for an effective transition to a circular economy. Synergies exist between the concept of digital twins and business models in Industry 4.0 due to their potential for innovation to shape business models (Adelsberger et al. 2024, pp. 409–414). They are used to simulate processes, map reality through real data and testing, and validate designed solutions, including in the field of the circular economy. The ability to transform these resources into useful semi-finished products, raw materials or even products creates conditions for acquiring new sources of sales, which in turn allows for the creation of new revenue generators. Through the circular economy, the green transition takes place, not only reducing the adverse impact on the environment, but consequently positively affecting the climate. In many cases, the green transition is supported by the digital transformation, which allows for the creation of business models and systems for monitoring environmental parameters such that it is possible not only to create new products and services, but to produce them in environmentally positive conditions. This model of operation treats waste as raw materials that have a useful value for their use in another form. The change from the linear economy to the circular economy leads to the transformation of business models of companies and thus defines them as green business models, i.e., those where the pro-ecological orientation is a factor stimulating value delivery to customers. In this cycle, it is necessary to reduce waste to a minimum and achieve a marginal level of waste generation in this respect. Extending the life cycle of products, but also sharing things, leasing to other entities, repair, refurbishing and recycling, but also reusing, are tools and mechanisms by means of which to achieve the status of a business model based on the circular economy. In such a model, a closed loop is created, where production and consumption are based on the minimum production of useless waste constituting economic losses. It should be noted that the circular economy requires the reconfiguration of the business model in such a way that the transaction takes place, i.e., the transformation of the entire system that will help achieve the so-called planned wear and tear. This means designing a business model where products are manufactured in a planned way such that it is possible to estimate their lifespan. It is possible to determine the time when they will cease to function as intended. This time should be as long as possible for the benefit of the environment, but also for climate change. The generation of additional value from useless waste improves the financial condition of companies, broadens the range of value delivered to customers and reduces the unit cost of producing a product or service. The condition is to find such technology that will allow the transformation of high-value raw waste materials to products or products delivered to customers. The reconfiguration of the business model in this direction is a complete change, because it is based on the idea of changes in the rules of consumption and production aimed at climate change. The transition to the circular economy is largely based on the assumptions of mechanisms of socio-technical systems. These activities are supported by digital platforms and are based on Big Data sets, where the effect is multiplied to many more units. Reducing pressure on the environment through monitoring systems based on digital solutions and embedded in business ecosystems around which large social groups are concentrated reduces pollution and other adverse environmental impacts. It helps monitor how the company produces products, what the social impact of these products or services is, what the level of carbon footprint is and what to do and when and under what circumstances to optimize this impact in a dynamically controlled system. Socio-technical systems understood in cybernetic terms based on modern technologies are able to change the internal layout of the production of products or services after obtaining feedback in order to minimize the adverse impact on the environment. The combined implementation of circular economy assumptions and cybernetics supporting the assumptions of the management process, including socio-technical systems, creates new space for building business models and designing products and services for changes in manufacturing technology. This approach is now a condition for companies gaining a competitive advantage and their widespread social acceptance.

6.7 PROGRESSIVE BUSINESS MODELS AND SOCIO-TECHNICAL SYSTEMS

A progressive business model (PBM) is a type of business model where legislative arrangements, i.e., legal regulations of assumptions aimed at implementing objectives conducive to improving climate conditions, are a factor stimulating the development of its mechanisms (Rochlin 2021). Progressive means focused on change. Progressive, i.e., following new trends, is the domain of this group of business models. Progress contributes to addressing not only the legal, but also non-legal challenges of society such as listening to people. Legislation that increasingly burdens creators with limitations requires innovative ideas on how to deliver value to customers and capture value from the market while fulfilling such stringent regulatory requirements. Progressivity is based on a model of strategic thinking that responds to new development opportunities. At the same time, it should be noted that new legislation, even if rigorous in relation to the impact on the environment, is a market opportunity for modern innovators. Managers who respond more quickly to these changes can become pioneers not only in creating value for customers, but also in setting new standards for service delivery. Adaptability to change is a trait of creative managers. Promoting eco-innovation under the conditions of legal changes is a progressive process carried out by means of small changes in the entire business model according to the principle that change in one place focuses on the whole of the changed object. These changes concern the modification of organizational structures, the redesign of processes, the withdrawal of certain raw materials and materials, the transformation of technologies in the pro-environmental direction and the digital transformation. The search is focused on finding alternative solutions, often using the assumptions of the digital economy. In this constellation of activities, a kind of new re-engineering of processes not only focused on economic aspects, but in particular on pro-climate aspects, is created. The search for substitutes for raw materials and semi-finished products as well as the search for new areas of activity determines new determinants of business activity. The disadvantage of these business models is that they arise from limitations through evolution. The strong interference of legal factors restricts not only business, but also the freedom of citizens.

According to the assumptions of strong legal regulation, especially in Europe with regard to climate change policy, the creation of business models requires a special innovative and creative approach to unlocking simultaneous economic opportunities while maintaining strict legal regulations. Mechanisms embedded in socio-technical systems are conducive to these solutions, as they mark progress in artificial intelligence, prediction, forecasting, monitoring, and the energy, green and digital transformation.

6.8 INCLUSIVE BUSINESS MODELS AND SOCIO-TECHNICAL SYSTEMS

Inclusive business models appear in numerous approaches to business models in the relevant literature, but they are also applied. Inclusion refers to the poor, who can also create economic value. The assumption of these models is to create value through the logic of creating, delivering and capturing value by providing access to resources such as finance, technology, distribution channels and creating conditions for decision making in this area (Rosenstock et al. 2020, pp. 76–81). Poverty is a driver of these models. Business in these models was created using elements that integrate new markets and new participants using the so-called hybrid value chain. In this approach, it is logical to seek synergy between economic and social value. It is also crucial to blur the boundaries between the commercial, social and public sectors and to create social innovation. In addition to market values, the public interest is highly visible in this approach. This is because it determines the contemporary factors responsible for the success or failure of companies. Without social market acceptance, business ventures are doomed to failure. Therefore, the core premise of this approach is the joint achievement of economic and social objectives without prejudice to any of them. This is particularly important in the so-called emerging markets, where poverty is high and economic growth is increasing. The principle is to include not only customers in the value chain, but also excluded people who can promote the creation of economic and social values to their benefit. It is not about treating them as a tool, but as a partner. Such a business is not meant to be a sale to the poor, but a joint venture. Examples of such business models could be, for example, micro-loans to the poor or setting the price for electricity according to consumers' purchasing power. The fundamental idea in this case is to see customers in groups characterized by poor income and even poverty. Moreover, the aim is to include them as partners in the value chain as distributors, suppliers, contractors, users, customers, etc. For example, middle-income earners can provide opportunities for poor people to participate in business based on their inclusion in the manufacturing chain. It is also both a principle and practice that stakeholders create an inclusive business ecosystem where appropriate value for each participant in the business creation chain is exchanged and delivered.

6.9 CONCLUSIONS

This chapter demonstrates that new ideas for creating business models are based on innovative and popular ideas that need to be accepted by specific groups of generations of people and are different for older people and younger people. Designers must pay attention to what is relevant to a given member of Generation X, Y or Z at a given moment, how to reach their expectations and how to satisfy their needs. Complex systems theory, game theory, general systems theory and other solutions fit into the need to professionally build new business space and shape innovations that would be impossible without theoretical knowledge and previous progress in science, engineering and management. The chapter proves that the condition for the emergence of new business solutions is non-standard thinking and challenging modern theories based on their knowledge.

REFERENCES

Adelsberger, R.T., Antons, O. and Arlinghaus J. 2024. Digital Twins and their Implications for Business Models: Overview and Potentials, IFAC PapersOnLine. 58–19.

Appelo, J. 2011. Management 3.0, Leading agile developers, developing agile leaders. Pearson Education, Inc. 30.

Austin, U., Olalekan, D., Babatunde, Y. and Okunade, L. 2023. The integration of artificial intelligence into database systems (AI DB integration review). International Journal on Cybernetics and Informatics (IJCI). 12(6): 161.

Barreira, L. and Valls, C. 2013. D Borgatti ynamical Systems. Universitext.

Borgatti, S.P. and Halgin, D.S. 2011. On Network Theory. Organization Science. 22(5): 1168–1181.

Chiang, R.H.L., Barron, T.M. and Storey, V.C. 2023. Data, knowledge and information in database and knowledge-based systems. Journal of Database Administration. 3(3): 18.

Cook, R.I. 2020–2021. How complex systems fail, from research to practice. Human Organizational Performance (HOP). Newsletter Psychological Safety.

Darwin, C. 1859. On the Origin of Species by Means of Natural Selection. John Murray, London.

Daryani, S.M. and Amini, A. 2016. Management and organizational complexity. Procedia—Social and Behavioral Sciences. 230: 359.

Elkington, J. 1997. Cannibals with Forks: The triple bottom line of 21st century business. Oxford: Capstone.

Frawley, W.J., Piatetsky-Shapiro, G. and Matheus, C.J. 1992. Knowledge discovery in databases: An overview. AI Magazine. 13(3): 58.

Govinndarajan, V. and Venkatraman V. 2024. Fusion Strategy: How Real-Time, Data and AI will Power the Industrial Future. Harvard Business Review Press.

Grabowski, M. and Wyciślak, S. 2015. Problem of contagion in complex systems. Jagiellonian Journal of Management. 1(2): 106.

Ireland, V. 2013. Exploration of complex system types. Procedia Computer Science. 20: 248.

Kellert, S.H. 1994. In the Wake of Chaos: Unpredictable Order in Dynamical Systems (Science & its Conceptual Foundations Series SCF). Chicago: University of Chicago Press.

Madanchian, M. 2024. The role of complex systems in predictive analytics for e-commerce innovations in business management. Systems. 12(415): 2.

Mynarski, S. 1981. Elementy teorii systemów i cybernetyki [Elements of Systems Theory and Cybernetics], PWN, Warsaw. 36–39.

Nunes Amaral, L.A. and Uzzi, B. 2007. Complex systems—A new paradigm for the integrative study of management, physical, and technological systems. Management Science. 53(7): 1034.

Raaidi, S., Bouhaddou, I. and Benghabrit, A. 2018. Is supply chain a complex system? MATEC Web of Conferences. 200(00018): 1.

Rochlin, C. 2021. Climate change and the progressive business model. The Electricity Journal. 34(4).

Rosenstock, T.S, Lubberink, R., Gondwe, S., Manyise, T. and Dentoni, D. 2020. Inclusive and adaptive business models for climate-smart value creation. Current Opinion in Environmental Sustainability. 42: 76–81.

Ross Ashby, W. 1957. An Introduction to Cybernetics. London, Chapman & Hall Ltd.

Thom, R. 1980. Paraboles et catastrophies. Saggiatore. Milan.

von Bertalanffy, L. 1929. Kritische Theorie der Formbildung (Translated by J.H. Woodger as Modern Theories of Development: An Introduction to Theoretical Biology. Oxford: Clarendon Press, 1933). Berlin: Gebruder Borntraeger.

von Neumann, J. and Morgenstern, O. 1953. Theory of Games and Economical Behavior. Princeton University Press.

Wiener, N. 1948. Cybernetics or control and communication in the animal and the machine. MIT University Press, Cambridge, MA. 212.

Zamiri, M. and Esmaeili, A. 2024. Methods and technologies for supporting knowledge sharing within learning communities: A systematic literature review. Administrative Sciences. 14(17): 1.

Chapter **7**

Assumptions and Results of Research into Socio-technical Systems in the Context of Designing Digital Business Models

7.1 INTRODUCTION

The research part of the monograph is devoted to the attempt to combine knowledge in the field of socio-technical systems with a modern approach to designing digital business models. The cognitive gap was the identification of a set of factors responsible for the effective creation and management of the development of companies, which are based on the assumptions of business models built on the use of digital transformation, where technological factors and business platformization shape the business environment based on the assumptions of the functioning of socio-technical systems. This approach has very innovative features, as it triggers a new perspective on the functioning of the digital economy. It explains why the combined implementation of social and technological factors by means of digital platforms has allowed key players in digital business to achieve great success and why many users are satisfied that they can use such advanced social and technological solutions thanks to the availability of mobile devices. The ability to design such solutions requires both theoretical and practical knowledge, and the

DOI: 10.1201/9781003456421-7

widespread availability of the Internet makes it possible to conceptualize and operationalize new business space. This chapter focuses on the development of original theoretical and practical models that should be used by researchers in the discipline of management science, but also managers who want to better understand the reasons for the success or failure of designing the operation of digital platforms. They want to get to know them not only superficially, but in such a way as to understand what their logic is and what management and engineering theories support these projects and what roles they play in them.

7.2 THEORETICAL ASSUMPTIONS OF RESEARCH INTO SOCIO-TECHNICAL SYSTEMS IN THE CONTEXT OF DIGITAL BUSINESS MODELS

The Main Objective of the Scientific Discussion

The main objective of the scientific discussion is to demonstrate the determinants of shaping digital business models and complex technological enterprises based on the assumptions of socio-technical systems theory. Current scientific achievements in this area allow for the use of these concepts in new conditions based on digital platforms and advanced technologies that fit into the assumptions of the network paradigm. The aim of the monograph is to show extremely different approaches to the construction of business models, where, despite their differences, the theoretical basis is a joint examination of social and technical factors. The aim is to show how modern technology currently makes it possible to build interrelated community networks. Thus, business ecosystems are created without borders, which, thanks to technology fusions, improves their performance in terms of efficiency and effectiveness.

Setting a Scientific Problem and Identifying a Cognitive Gap

As part of the scientific goal, it is crucial to determine the scientific problem and identify the cognitive gap. This cognitive gap is an identifiable set of factors responsible for the effective creation and management of the development of companies, which are based on the assumptions of business models built on the use of digital transformation, where technological factors and business platformization shape the business environment based on the assumptions of the functioning of socio-technical systems. Linking the features of socio-technical systems assumed by designers and business planners helps build a business ecosystem that allows business subsystems to connect with larger systems, creating a global market for the exchange of information and values.

Such a defined cognitive scope should lead to the emergence of a new concept of designing digital business models which are suitable not only for innovative solutions in the information and communication sphere, but also in traditional forms of doing business in the railway, energy or water supply sectors, where technical infrastructure determines the conditions for delivering value to customers, and a community is built alongside it.

This concept should fit into the current conditions of business creation based on the latest digital technologies and an approach to the development of business ecosystems.

Set against the background of this concept, theoretical and practical factors determining the high efficiency of the functioning of companies supported by the processes of digital transformation based on the use of modern technologies conducive to building communities and the exchange of values between them, and even the interdisciplinary, dynamic migration of values, should emerge.

The author understands that engineering, technological solutions and digital transformation break the current paradigms of strategic and operational management and should be explained again by means of different assumptions and principles of describing modern business, including its limitations. There is a need to combine social and technical aspects in parallel to explain phenomena in the area of new theories and build utilitarian solutions. Leaving aside engineering aspects and relying on previously known and well described management paradigms is a mistake and leads nowhere. What has been recognized in the management sciences as dogma before the widespread implementation of digital solutions, including artificial intelligence, is of little value today. New paradigms need to be sought in the interdisciplinary use of social and technical sciences. In the context of complex technical objects, whether based on purely IT solutions or based on the assembly of technical infrastructure and organizational elements in order to achieve the expected results in the sphere of functions and performance of socio-technical systems, it is crucial to design such solutions that must be based on other mechanisms. These mechanisms must combine the achievements of social sciences and technical sciences.

According to the author, the presentation of the issue in this way requires defining new assumptions for the theory of management and development and the practice of business models with a strong use of technological solutions conducive to the improvement of dynamic communication between various groups of stakeholders focused on the perception of the value generated by these business models.

Background to the Scientific Discussion

As part of the adopted theoretical assumptions related to the solution of a defined scientific problem associated with the management of the efficiency of railway sidings, an analysis was conducted to embed this problem in the theory of management sciences.

In order to correctly identify the phenomenon in science, it is important to refer to scientific discussions that determine the principles of the adopted, appropriate interpretation. To this end, however, it is necessary to combine the conditions for understanding the assumptions of the digital economy and digital transformation compared to the efficiency of complex technical objects that create conditions for the delivery of value based on the necessary technical infrastructure. The effectiveness of a complex technical facility will depend on the operational condition of the facility, i.e., its condition as defined by a set of values of its technical and economic characteristics determined for the facility at a given moment or within a specified period of time. Value is created when this facility can go beyond its borders through interfaces that allow it to be connected to other technical systems in this way, while respecting the principles of technical and functional safety and cybersecurity. Such systems can create added value by creating a complex business ecosystem and, through the synergy of the socio-technical system, create added value for many market participants. It is important to combine the effectiveness of the organization in terms of the effectiveness of its management system with the effectiveness of a complex technical object. The condition is to apply the approach of operational reliability and transfer of information and knowledge between the entities included in the system and the object of the services provided or production.

The Adopted Research Methodology and the Scope of Research and Scientific Discussion

In the area of research and scientific discussion, a systematic review of the literature on the concepts of the socio-technical system in the business model was conducted. A critical review of the literature was conducted, which underlay a number of the author's original perspectives of the topic at hand. Several case studies were described, both in the field of "purely" digital business, e.g., Google, and the analysis of case studies for the railway, energy and water and sewage sectors. The latter are in line with the assumptions of socio-technical systems theory and undergo dynamic transformational processes in the area of digitalization, not only of individual processes, but of the entire system of action that changes their business models, often occurring in the formula of the multi business model environment.

In order to achieve the assumed goal of the monograph, qualitative theoretical and design research was used to lay the theoretical and practical foundations for the design and development of digital business models based on the assumptions of socio-technical systems theory.

The research method identified activities to solve a given scientific problem, which is presented in Table 7.1.

Table 7.1 Sequence of Activities within the Research Method used to Solve the Scientific Problem of the Place and Role of Socio-technical Systems in Shaping and Developing Digital Business Models.

No.	Sequence of activities of the research method of mechanisms for the effective management of railway sidings
1.	Collecting relevant literature on the subject of research into the place and role of socio-technical systems in shaping and developing digital business models.
2.	Identifying factors explaining the principles of designing digital business models in the context of social and technical determinants.
3.	Defining the extent of the influence of various factors on the design and development of complex socio-technical systems that create business ecosystems for the functioning of digital business models.
4.	Selecting a research sample to investigate the mechanisms of using socio-technical systems theory to build conditions for the development of complex business solutions supported by the digital economy.
5.	Defining the characteristics and determinants of complex economic systems operating under the principles of the digital economy.
6.	Analyzing contextual conditions and shaping different business models based on different theoretical and practical assumptions.
7.	Defining good practices in the use of socio-technical systems and developing a theoretical model for the description of this system to create a theory of designing digital business models operating within the framework of systems complexity theory.
8.	Developing analyses, the resulting synthesis, conclusions and recommendations for the improvement of business based on digital solutions in the environment of many opposing and mutually synergistic technical systems.
9.	Determining the theoretical determinants of creating innovative digital business solutions that build the business model as a whole and support their selected functions, often based on technology fusions.
10.	Identifying new directions for the development of scalable digital business models embedded in the complexity of socio-technical systems.

Source: Own study.

Such defined scientific research allowed for the formulation of research questions and the achievement of the goals defined within the scope of this monograph.

The author's extensive experience as both a scientific researcher and a business practitioner allowed him to both create a research sample frame and conduct extensive theoretical and practical research. The study is a cognitive–evaluating study based on multidimensional scientific inference. The monograph covers both the assumptions of creating a new theory of shaping and developing digital business models, based on the assumptions of business ecosystems created based on the conditions of socio-technical systems, and elements including practical solutions that can be applied.

The Subject of the Scientific Discussion

The subject of the scientific discussion, research and analysis includes scientific and application digressions on components forming a coherent system of joint examination of the processes of designing innovative business models, the use of technical systems for the development of assumptions for the ecosystem that is part of a larger whole and technology fusions creating complex business solutions.

Cognitive Objectives

Cognitive objectives include:

1. Identifying the factors responsible for shaping socio-technical systems.
2. Determining the relationship between the processes of designing digital business models based on digital platforms or technical infrastructure systems built from a network of links and the theoretical attributes of socio-technical systems.
3. Assessing the impact of social and ecological factors on the development of unique attributes of digital business models supported by popular social ideas.
4. Assessing the relationship between the conceptualization and operationalization of an innovative business idea through different theories.
5. Developing a new paradigm of strategic management that counters the theories unsuitable for the modern economy with concepts that do not fully work in the digital and global environment.

Methodological Objectives

Methodological objectives include:

1. Identifying the factors responsible for ensuring coherence between socio-technical systems theory and the requirements for the design of digital business models and the digital transformation of technological companies.

2. Developing strategic recommendations for the process of scalability of solutions in the sphere of digitalization and taking into account different types and concepts of business models in terms of social impact thanks to technologies.
3. Developing a universal description of technological platforms grouping communities for the creation of economic value in the form of a business model (canvas).
4. Developing taxonomies of digital business models based on the assumptions of socio-technical systems.

Utilitarian Objectives

Utilitarian objectives include:

1. Developing an original methodology for designing digital business models and technical systems undergoing digital transformation.
2. Developing a set of management and technical–operational factors with the greatest impact on the effectiveness of business models of companies, the operation of which is conditioned by technologies and the role of large communities and Big Data sets.

Key Research Questions

The following research questions were posed within the framework of the defined research objectives:

1. Is there a common denominator in the design of digital business models, where the genesis of this process is the use of the assumptions of socio-technical systems theory?
2. Is there synergy between the type of ideological approach adopted to define the social and managerial functions of the business model and its location in the digital economy in this network ecosystem?
3. Is the network paradigm conducive to building complex systems built from technical infrastructure and how can one design interfaces between business solutions of different companies not located in one sector of the economy?
4. What set of factors shapes the possibilities of entering into technology fusions that build a networked, digital business model embedded in specific legal, economic, social, political and ecological realities?
5. What methods in the field of management, what aspects of creating innovations, including open innovations and

technology fusions, and what principles of building network relations between systems and subsystems forming the business ecosystem are important for designing digital business models?

Embedding the Topic in Scientific Theories:

Due to the multidimensionality and holistic nature of the presented content, the topic of the scientific discussion requires one to address issues related to the theory of management and quality science, economics and finance, and the theory of technical sciences.

This Book is Based on the Following Assumptions:

Macroeconomic assumptions

1. When it comes to digital solutions, they are very often global, even appearing in the Open Source formula, and project cooperation can involve knowledge workers from different parts of the world thanks to the Internet.
2. IT solutions for creating business solutions are universal and generally available.
3. Global social, environmental, political, military, economic and financial crises adversely affect international cooperation and should be taken into account in decision making pertaining to the implementation of economic ventures in the form of projects.

Sectoral assumptions

1. Economic sectors, including high-performance and high-reliability organizations such as railway, energy and water and sewage companies, but also all others based on a combination of technical and organizational issues, are undergoing dynamic transformation processes towards the challenges of the digital economy, which generates the creation of new business models that break the previous rules in the sector, fulfilling the principles of the concept of a "game changer".
2. The professionalization of the industry is crucial, and the expectations of societies force the creation of social enterprises, where, in addition to economic profit, it is important to generate social profit.
3. The standardization of business model design processes requires consideration of the legal conditions relevant to the functioning of companies embedded in specific sectors of the economy. Legislative issues should be taken into account as well.

4. Companies operating in the sectors are regulated by law in terms of their activities in numerous areas. The socio-technical systems where they operate are often subject to control by authorized supervisory authorities, which affects quality, safety and efficiency and equal rights vis-à-vis the market. This has a significant impact on the determinants of designing digital business models within the constraints of the sector in which a given company operates.

Microeconomic assumptions

1. A great number of companies relying on the extensive use of technical infrastructure in the *operator–technical formation–working environment system* are not fully aware that they operate in a socio-technical system. A lack of such knowledge creates a number of problems when implementing digital transformation projects.
2. The business environment is forcing changes in the business models of companies, particularly in terms of better availability of services through digitalization, enhanced communication and access to information. Massive data based on Big Data, including customer and potential customer data, is used to create large volumes of community around a business model.
3. Modern digital business models and solutions that transform analog business to digital business based on the theoretical assumptions of socio-technical systems supported by IT/technological solutions require receipt of a set of guidelines on how to conduct the process so as to maintain business continuity alongside a dynamic process of revolutionary changes adapting to new conditions of doing business.

This book is based on the following limitations:

1. So far, there has been a lack of extensive research into the place and role of socio-technical systems in shaping digital business models, and such business models that need digital transformation for companies that operate on the basis of combining technical infrastructure with work organization and the huge role played by the human factor. There is not much experience in this area, including scientific publications.
2. The multitude of approaches to naming and codifying the names of business models prevents the establishment of one binding hierarchy (taxonomy) in this regard.
3. The presented examples of a unified description of business models based on the author's original approach to the determination of

variables which describe socio-technical systems in the context of the digitalization criterion characterize different sectors of the economy, which on the one hand, is a limitation, but the unification of such a theoretical and practical description can also be valuable for other companies from other sectors of the economy.

Time Relevance, Importance of Issues and Contribution to Science of Issues Related to Socio-technical Systems in Shaping and Developing Digital Business Models

By undertaking the writing of this monograph, the author wanted to bring added value to the identification, conceptualization and operationalization of the implementation of mechanisms for the use of the assumptions of socio-technical systems for the design and development of digital business models – digital platforms and technical infrastructure facilities in technological enterprises.

The proposed problem does not exist in the relevant literature in such broad terms. The value of the creative contribution is related to the presentation of a comprehensive approach to the types of digital business models, the unification of their description, and the inclusion of artificial intelligence, a network paradigm, business ecosystems and social impact, safety and cybersecurity in the creation of innovative business solutions in this area. Operational aspects have also been taken into account within the framework of process excellence.

7.3 QUALITATIVE RESEARCH AND CASE STUDIES

The main objective of the scientific research in this monograph is to demonstrate that socio-technical systems as a theoretical and utilitarian concept as regards digital platforms and complex industrial systems such as the energy, railway, water and sewage sectors, etc., which are currently the fundamental formula for the design and development of digital business models and digital solutions as part of digital transformation processes, constitute a core theoretical and practical resource for the design of modern solutions in this area. The book includes extensive literature studies based on a systematic and critical literature review and modeling based on the analysis of four case studies, namely:

1. A case study of a complex system based on the assumptions of the digital economy, namely multi-sided digital platforms.
2. A case study of a complex industrial system, namely the rail transport sector, where the role of the operator (a human,

technical system and working environment) specific to the era of implementation of digital transformation solutions is changing the face of this sector, creating a set of digital business models supported by digital functionalities cooperating within the assumptions of the socio-technical system. A special role in this system is also played by the role of technical safety and cybersecurity, which is important for understanding the dependencies in this complex system.

3. A case study of the water and sewage sector, which, like the railway sector, is composed of social and technical layers, which work together to deliver the expected results in terms of business continuity and in a concurrent analysis of the role of the human factor and technology. In this sector, due to the challenges related to climate change policy, modern business models are being created based on the assumptions of knowledge transfer, innovation and implementation of digital solutions on many levels of its operation.
4. A case study of the energy sector, where, especially in terms of the emergence of new business models of renewable energy sources and self-sufficient energy clusters, new digital business models are being created which are not only related to one or several functions, but those that integrate the functioning of the entire energy sector with the participation of prosumers and renewable energy solutions together with event monitoring and evaluation systems of variables measured in real mode.

In addition, the author's original solution was proposed in the sphere of modeling such systems and their description by means of solutions from the sphere of theory and practice of the concept of socio-technical systems. This approach can be useful for theorists and practitioners to better understand the construction of digital business models based on the assumptions of the concept of socio-technical systems.

Finally, the author also developed an original canvas of the description of digital business models and complex socio-technical systems, which can be used to design and characterize such specific forms of business activity stimulated by digital solutions and based on technological progress.

7.3.1 Socio-technical Systems—Case Studies for Designing Digital Platforms as Business Models

Today, digital platforms dominate modern business. The digital economy stimulating digital transformation processes in virtually all parts of the economy has created an avalanche of solutions combining technologies with the integration of large communities of communities. Even mass and multidimensional communication between users emerged that takes place on specialized e-commerce or social media platforms. Digital platforms meet the definitions of socio-technical systems because they combine core features of the role of technology and community participation. Technology fusions between different digital platforms, including the use of their functions and performance for new innovative solutions, build complex digital systems available in almost every part of the world via mobile devices and computers. The big four companies, namely Amazon, Apple, Facebook and Google, dominate this market. Using the theory of socio-technical systems to explain crucial success factors, but also the logic of business models and thus the way in which revenues are converted into profits, will allow for a better understanding of the successes and failures of digital platforms in the modern, dynamically changing world.

On the basis of standardized questions related to the assessment of socio-technical systems, the ecosystem of multi-sided digital platforms was described. To this end, the questions developed by Appelbaum (Appelbaum 1997, pp. 452–463) were used.

7.3.2 Socio-technical Systems as a Platform for Designing a Railway System—A Network Approach

In order to identify the characteristics of the rail transport ecosystem as a business ecosystem based on the assumptions of the socio-technical system, several important features have been identified. Transport systems in general, and railway systems in particular, are large and geographically dispersed systems. While individual carriers can be independently operated and managed, they are forced to use the infrastructure and resources (roads, tracks and stations) of other entities (Hoehne 2016, p. 36). As part of the description of the railway system, it is necessary to refer to the assumptions of the System of Systems concept. In subjective terms, a railway system is built from a structure comprising railway undertakings providing transport services for the transport of people or goods and infrastructure managers who provide railway lines with accompanying infrastructure to service railway undertakings. In objective terms, the infrastructure manager is equipped with a complex system of equipment and railway infrastructure used to control traffic in accordance with certain provisions. The railway undertaking, on the other hand, provides transport services in accordance with the rules laid down by the infrastructure manager. The infrastructure manager decides which railway track the railway undertaking will enter and how the train will be handled in terms of rail traffic management. Focusing on the infrastructure manager whose task is to ensure the safety of rail traffic when making railway lines available to railway undertakings, it should be noted that its business model is based on the concept of a system of systems. The overall railway system of the infrastructure manager is subordinated to the implementation of the train timetable, taking into account restrictions, safety criteria and ensuring a certain capacity of rail traffic. The railway system of the infrastructure manager fulfills the requirements of the socio-technical system because it has activities built into its structure resulting directly from the functions and technical performance of the equipment, as well as the key role of operators such as rail traffic controller and dispatchers. The socio-technical system subordinated to the operation of the railway system is a dynamic, complex, interactive and supervised system, where the technical and organizational solutions necessary for ensuring the continuity of rail traffic are introduced. The efficiency of such a system is maintained, provided that all other subsystems function properly and in harmony with the intended plan. The overlapping of the tasks of railway undertakings and the infrastructure manager requires responsibility for the tasks performed to be ensured. The human factor of the game plays a crucial role in the railway system, as they make a decision, execute it and take responsibility for it. All this happens in a limited time depending on the speed of trains, the intensity of rail traffic and the capacity of its individual sections. The technical layer consists of several structural subsystems, namely a structural control–command and signaling subsystem, a structural energy subsystem, a structural infrastructure subsystem, as well as functional subsystems such as rail operation and maintenance. All these subsystems must ensure the technical compatibility of the railway system with the technical specifications for interoperability and national regulations. Moreover, safe integration between them must be ensured. Safe integration must also be maintained between the rolling stock structural subsystem and the energy, control–command and infrastructure structural subsystems. The static level revealing itself in the form of the construction of the railway system, its architecture of the control–command and signaling equipment used must operate under the conditions of railway operation under specified constraints and in weather conditions. Harmonization of work processes, equipment operation and communication between all rail traffic participants is a condition for the safe use of this system. The socio-technical system in question implements innovations and different business models provided that they are in line with the technical compatibility and safe integration of the railway system. The railway system meets the definition of a complex system because it consists of a large number of elements that together ensure achievement as a function of the objective. Any deviation from the assumed parameter of a smaller system in relation to a larger one, in other words subordinate to the superior one, may cause system dysfunctions and, consequently, catastrophe in land traffic. The railway system is built by means of human–machine interfaces in the railway system that interact with each other in real time. Humans are somehow part of the technical system, because they perform core tasks therein and are its weakest link in terms of traffic safety. Therefore, they must be under constant supervision and constantly undergo training, instruction and authorization for specific jobs. The digitalization of the railway system is a concept embedded in its structure. Data is transmitted on secured frequencies allocated only to railway systems. The digital transformation creates a number of new conditions for the operation of these systems, in particular with regard to cybersecurity risks and the adverse impact of parties outside the railway system. The railway system is therefore a closed system which no outsider has access to. It is a separate system functioning on the basis of specific regulations, standards and norms. In subjective terms, the so-called unbundling principle prevails, whereby the roles of the railway undertaking and the infrastructure manager cannot be combined. This results from European Union regulations and aims to indicate unequivocal responsibility for the tasks performed as part of transport work and processes of making railway lines available. Solutions related to the digitalization of railways, including the use of digital twins, simulators, and digital data transmission, place railway undertakings as those that directly fulfill the requirements of the socio-technical system. An additional factor that influences this is the mass of recipients and the large number of employees involved in operational processes.

On the basis of standardized questions related to the assessment of socio-technical systems, the characteristics of the rail transport ecosystem in which railway undertakings work were listed. To this end, the questions developed by Appelbaum (Appelbaum 1997, pp. 452–463) were used what shows Table 7.2.

7.3.3 Socio-technical Systems as a Platform for Designing Water and Sewage Sector Companies

The water and sewage sector is of strategic importance for the development of local markets. In general, it functions as a natural monopoly and, through its companies, supplies water to collect sewage from a specific area. Water supply companies usually operate within cities or municipalities. They consist of a complex sewerage system in the form of a network of piping systems. A simple structure in a water supply and sewage company consists of top management, operational management and employees directly performing the processes of operation and maintenance of the sewerage network. Water and sewage

Table 7.2 Parameters Describing the System of Socio-multi-sided Digital Platforms

No.	**Parameters describing the system of socio-multi-sided digital platforms**
	1. Environment:
1. What are the major factors in the organization's environment that influences the work?	A digital platform is primarily a digital infrastructure that connects producers, consumers, service providers, advertisers and other groups in an interactive ecosystem and enables the exchange of goods, services or information. The relationship between humans and "non-human actors" in the "micropayment space" –profiting from the interaction between people, organizations and things.
2. What characteristic environmental problems arise?	Digital platforms are demolishing the current set of rules in business. Digital platforms have changed the business landscape in the global economy. Companies such as Uber—a company providing passenger transport services, which has no vehicles, and therefore uses both resources. The same applies to platforms such as Facebook, Google, Alibaba (retail) and Airbnb (hotels).
3. In which elements of the system are the interactions with the environment most visible?	Digital platforms are a "carrier", an environment of social "networks", possibly an indifferent "medium" providing services to society as a whole. Focusing on the potential of clouds may imply clean services and is not conducive to the examination of down-to-earth consequences. This applies to the energy required to power large server farms and their impact on the environment. Storing data on external servers rather than on a computer means that Facebook and Instagram files, for example, consume growing energy resources.
4. What is the organization's greatest strength *vis-à-vis* the environment (competitive advantage, labor market, productivity, etc.)?	The narrative power of platforms would not exist without the uniqueness of "power technologies", i.e., platform algorithms, especially social algorithms, that have direct contact with humans.
5. What is the organization's most obvious weakness?	The speed of technological progress results in the need for continuous modification and improvement of platforms. Solutions that completely change the rules of the game, so-called "game changers" may appear, which can change the game conditions on the basis of which digital platforms operate. New technologies can push older technologies out of the market.

No.	Parameters describing the system of socio-multi-sided digital platforms
	2. Goals/tasks:
1. What factors within the system most influence goal attainment (positively/negatively)?	The power of platforms is shaped primarily by the potential to impose interpretation + missionary rhetoric + opportunities to influence the social, political and economic environment, taking into account unique competences and the collection and use of resources, especially intangible ones. Negative qualities result from the possibility of manipulating the opinions of recipients and imposing a certain way of thinking on them.
2. Which elements of the system are most affected by changes in goals and tasks?	Platforms are constantly evolving under the influence of the political, legal or social environment and the development of programming knowledge. Another fundamental feature of platforms is crucial, namely its "interpretative performativity", i.e., the variability of what users and millions of companies see in the platform depending on which of its types of roles comes to the fore.
3. Are the rest of the elements coordinated with respect to the goals/objectives and tasks?	The tasks of digital platforms are embedded in open systems, constituting a common product of technology, social and physical structure and culture immersed in the environment it co-shapes – it is a network of collective activities undertaken as an attempt to shape the world and people's lives.
4. Where are the strengths and weaknesses in the interactions among elements?	The strength is that the digital platform shapes its own ecosystem, controls it, and decides how the world should look in this virtual ecosystem. The weaknesses are technological advances threatening their existence, the high costs of ensuring the efficiency of the whole system and legal issues arising from the place and role of platforms in the socio-political system.
5. What are the sensitive points that repeatedly cause friction, problems, or conflicts?	Conflicts occur in the context of the impact of digital platforms on the functioning of traditional business activities, for example Uber on the taxi market, Airbnb on the hotel market, etc. This gives rise to conflicts at legal and social levels.
	3. Structures:
1. Are the existing structures useful for the attainment of the goals and tasks?	Digital platform structures are focused on the creation of mass intellectual power – the power of digital platforms is the ability to impose will, force specific behavior and control the actions of humans (the user), social groups and economic entities.

(Contd.)

No.	Parameters describing the system of socio-multi-sided digital platforms
2. How do people adapt to or cope with these structures?	The human being, rhetorically objectified and in fact "commodified" on the platform, acting in several overlapping roles on the platform – as a consumer and producer and as a passive recipient, rarely as a citizen – in each of them contributes to the value of the platform, always as a data provider, sometimes as a co-creator of content and/or as a consumer.
3. Are the structures supportive of the needs and capabilities of the leadership (management) and the workforce?	Digital platform workers work in the field of design orientation. It is a multi-project environment. Each activity is coordinated by specific technological and organizational solutions based on a project approach, often involving an agile project management methodology.
4. Do the structures encourage co-operation and collaboration?	Algorithms of digital platforms are designed to enable interaction among all users. This is how these platforms work.
5. Do the structures facilitate a rational use of resources and technologies?	The level of optimization of resource use processes is very high. Systems are designed to achieve the best performance in terms of effectiveness and efficiency. It is important to aggressively acquire new users and maintain existing ones.
6. What major strengths/weaknesses result from the structures interacting with other elements in the system?	Digital platforms merge with other platforms to create a global system of interlinking multidimensional interfaces to enhance new features, performance, and the extent of the platform's impact on society.
4. Leadership procedures:	
1. Are the existing leadership procedures and processes focused on the specific goals/objectives and tasks?	Algorithmic authority is a type of leadership created in the work of digital platforms which gives more confidence in the results of the algorithmic process than in human knowledge and intuition. This gives recognition as an authoritative process of acquiring value from various, sometimes unreliable sources, without the participation of human expert evaluation. This is new leadership based on technology.
2. Do these procedures fit the organization's specific social and technical realities (technologies, other procedures, composition of the workforce, etc.)?	The procedures result from highly sophisticated algorithms for managing relationship networks in digital platforms. It is a modern system for monitoring processes and activities inside the system and its interaction with other systems based on the assumptions of the digital economy.

No.	Parameters describing the system of socio-multi-sided digital platforms
3. Are managers sufficiently trained in using the procedures (e.g., performance evaluation)?	Managers are very familiar with the rules of functioning of digital platforms and so steer their processes to achieve satisfactory results for the platform, build a competitive advantage over others and shape new space for enlarging the number of supporters of these platforms using an intentional approach to building development strategies.
4. What influences and changes from the outside are expected? What re-examination of the present procedures might these changes entail?	Changes in the functioning of digital platforms take place dynamically on the basis of cybernetic assumptions through the use of simple couplings and feedback loops, the results of which are directly implemented for the needs of new versions of the application. Change is a fundamental part of the management of these platforms; they are constantly updated and adapted to new needs. This is controlled by self-learning algorithms using deep learning and artificial intelligence.
5. What are the most significant strengths and weaknesses in the leadership procedures and their application?	The strength of leadership is that technology, mobile devices, computers, servers, etc. in the technical sphere are built only to shape systems of communication between users and customers in the case of monetization in order to build these active links between society and the strategic assumption of the functioning of digital platforms.
	5. People:
1. What is the basic attitude of various employee groups towards the organization?	Employees in corporations carrying out activities related to the construction of digital platforms work on the usual principles based on the idea of a design approach. In these types of organizations, the form of work is often based on the so-called turquoise organizations and are result-oriented according to the concept of performance management.
2. What are the effects of these attitudes on performance and teamwork?	The efficiency of this type of organization is very high, because it is based on the high competence of the staff and the selection of members to the project team. Often teams are built on the principles of diversity in order to create different solutions due to different experiences of the project team. Young and old, experienced and inexperienced employees mean that, based on the principle of brainstorming, very innovative creative solutions can be developed.

(Contd.)

No.	Parameters describing the system of socio-multi-sided digital platforms
3. How do managers and their subordinates adapt to the existing structures?	As already mentioned, work in this type of organization is based on creating effective and efficient result-oriented project teams. Although the work is carried out under pressure, the organizational culture is to implement difficult problems with the support of other project team members.
4. Do the employees feel that the structures, technologies and procedures serve as an incentive to full involvement (motivational effectiveness)?	Motivational effectiveness is important when employees are focused on achieving good results, especially in the context of the fact that time pressure is so high. Then, every mistake that affects the dysfunctions of the process system and errors in projects affect the outcomes of project review verification and validation.
5. Where are the biggest obstacles, problem areas and sources of conflict?	The biggest problems are time, pressure on results, maintaining high competences, communication obstacles and pressure to create innovations. The sources of conflict are neutralized by good management, but they cannot be avoided due to the specificity of human nature. Many activities are now supported by means of artificial intelligence, which reduces problems in communication between people and serves to eliminate barriers resulting from mismatches between project teams.
6. Technologies:	
1. How do changes in procedures and support systems affect people and structures in the organization?	Procedures are based on algorithms of embedded digital platforms underlying the functions and performance of these platforms. People are focused on creating new algorithms, improving current ones and fitting into the scope of innovative solutions with their creativity, organizational skills and talent management.
2. Do changes produce resistance?	In any organization, change generates resistance. Each change has a specific cycle that managers must explain and implement solutions conducive to conducting these changes.
3. When technological changes occur, are the structures and processes adapted or modified accordingly?	Every change in an organization operating in the digital economy, especially given the specificity of project management, involves changes in structures and procedures that must be adapted to new conditions. Technological changes entail a whole series of adaptive processes aimed at transforming the organization into new development directions in relation to new requirements that have been developed within the framework of new innovative technologies.

No.	Parameters describing the system of socio-multi-sided digital platforms
4. Is there a timely attempt to prepare people for technological and organizational changes, e.g., through personnel management, training or participatory decision making?	Modern organizations have specific structures responsible for training, mentoring, coaching and providing the necessary competencies to ensure the engagement of employees who are able to quickly adapt to changes and introduce them with a high level of efficiency.
5. What and where are the most important strengths and weaknesses between technology (including processes and procedures) and the other elements of the system?	Modern technologies stimulate changes in procedures and processes. They dominate the standardization and unification of processes. New technologies determine changes therein and are a priority because they respond to the needs of customers and other stakeholder groups. Standardization itself in high-creativity organizations should keep pace with changes to achieve operational excellence. Thus, changes occur from a startup to a business model, assuming that the startup does not generate operating revenues while the business model has the ability to generate revenues from the sale of products or services.

Source: Own study based on Appelbaum (1997, pp. 452–463) and Kreft (2021, p. 25).

systems fulfill the requirements of socio-technical systems because the key role in supervising the maintenance of the efficiency of the water sewage system is played by employees. They are closely connected to the network, working on its reliability, supporting processes related to the supply of water and ensuring its purity, availability and collection of sewage treatment in accordance with environmental regulations. The relationship between a social factor in the form of a team of employees assessed by external stakeholders works to ensure the efficiency of the operation of the water and sewage system, which is a critical element of this system. A technical and social issue combine to ensure the continuity of the system. In view of the challenges faced by new climate-changing business models, the prospect of a new energy sector based on renewable energy sources and the savings required in operating activities, these organizations use a number of innovative solutions that reduce costs and increase efficiency. The multidimensional arrangement of links between technical and organizational factors forces managers to constantly monitor operational processes and ensure the efficiency of technical infrastructure. The network of water consumers and sewage suppliers expects the technical system to operate reliably. The technical system is also subject to numerous threats resulting from potential terrorist or sabotage activities, especially in the situation of many political, social, economic and military crises. The assumptions of the digital transformation of the water and sewage sector are based on the implementation of IT solutions not only at the level of individual processes, but in the scope of the entire system, taking into account circular economy solutions. The implementation of waste-free regulations is shaping new space for the creation of digital business models based on innovation. The human and organizational factor supported by digital transformation creates new solutions in the sphere of operational excellence by combining the social factor with the technological factor. A great number of customers classify companies in the water and sewage sector as organizations based on the assumptions of the socio-technical system.

Based on standardized questions related to the assessment of socio-technical systems, the characteristics of the ecosystem of water and sewage companies were developed. To this end, the questions developed by Appelbaum (Appelbaum 1997, pp. 452–463) were used what shows Table 7.3. and 7.4.

7.3.4 Socio-technical Systems as a Platform for Designing Power System Companies in the Energy Sector

The energy sector is of crucial and strategic importance to the development of the country. Energy companies have extensive infrastructure ranging from power plants to high-voltage grids, low-voltage grids and distribution system operators, supplying electricity to institutional and individual customers. Energy companies are characterized by complexity and fit into the concept of socio-technical systems. Like the customers of rail and water services, they have a mass customer. They operate in a distributed, networked system and the technical infrastructure allows access to remote points. There is an extensive technological structure composed of many elements that together form a uniform system. The generation and distribution of energy creates a modern business ecosystem which, on the one hand, is powered by traditional sources of energy generation and, on the other hand, is supported by renewable energy sources. In the global economy, the percentage of renewable energy sources is growing, which is strongly related to the creation of modern business models affecting positive climate change. Caring for the natural environment requires the creation of new innovative business models in the energy sector. The assumptions of the socio-technical system have become a platform for building such business models. The digital transformation of circular economy solutions is also conducive to the creation of technological innovations supported by well-structured organizational models. Technologies are critical to building a network of relationships to meet the demands of different stakeholder groups. As part of technological progress in the energy sector, the paradigm in energy production shifts from a centralized model to a decentralized model with the participation of renewable energy sources. This evokes the need to create solutions based on prosumers, who, on the one hand, generate energy and, on the other hand, receive it. In this way, the end customer creates the conditions for developing new business models. Energy clusters emerge to create self-sufficient energy systems operating on local markets. Opening the network to distributed sources allows for the use of accumulation capacities and the creation of innovations. Small-scale energy generation is also a very interesting direction of development that shapes new space for building new business models. Microgeneration and prosumer-based solutions are changing the rules of the game in the sector. Smart electricity grids allow for better communication among energy market participants, cost reduction and increased efficiency, as well as integration of distributed energy sources, including from renewable energy. The digitalization of monitoring systems which help measure energy consumption and obtain information allows for the optimization of costs and elimination of losses in the network. Technology is therefore a stimulus for creating new innovative business models, taking into account the place and role of people as well as communities, which significantly affect the effectiveness of power systems. The energy sector understood in this way fits into the concept of socio-technical systems and allows for

Table 7.3 Parameters Describing the Socio-technical System for the Ecosystem of Railway Transport Companies

No.	Parameters describing the socio-technical system for the ecosystem of railway transport companies
	1. Environment:
1. What are the major factors in the organization's environment that influences the work?	The main factors influencing work are time pressure, which is based on the need to perform work activities at a specific time following other rail transport operations. The work is carried out with a high level of responsibility for the decisions made, for which there is not much time to think. Most decisions are made in accordance with specific operating procedures.
2. What characteristic environmental problems arise?	These are weather conditions, various disruptions affecting the deviation of the process from the plan (e.g., faults, delays, failures, accidents, incidents, operational errors, poor planning, bad reactions to events, etc.)
3. In which elements of the system are the interactions with the environment most visible?	The railway system is permanently built in the field and has the character of a linear investment. It affects its environment not only by affecting natural resources, but also the environment of human life because the density of the railway network is such that almost every citizen has contact with it. It directly interferes with the natural ecosystem and the ecosystem of human life; it is an indispensable part of it.
4. What is the organization's greatest strength *vis-à-vis* the environment (competitive advantage, labor market, productivity, etc.)?	The railway system is a natural monopoly. With regard to the activities of the infrastructure manager, with a few exceptions representing a small percentage of the entire rail network, it constitutes a monopoly. It is therefore subject to a number of specific legal provisions ensuring unrestricted access to the railway infrastructure of other entities in order to ensure the competitiveness of rail transport vis-à-vis other modes of transport.
5. What is the organization's most obvious weakness?	The biggest weakness of the organization is that it is a so-called high reliability organization, i.e., the probability of accidents and catastrophes is very low, but their consequences, once they occur, are very serious (e.g., death, loss of health, significant material losses, adverse impact on the environment).

No.	**Parameters describing the socio-technical system for the ecosystem of railway transport companies**
	2. Goals/tasks:
1. What factors within the system most influence goal attainment (positively/negatively)?	The degree of goal attainment is most influenced by the efficiency of operational processes as well as resource efficiency and effectiveness. A positive organizational culture, including a safety culture, should have a positive impact. In particular, poor technical infrastructure, insufficient training and a lack of strong employee motivation can have a negative impact.
2. Which elements of the system are most affected by changes in goals and tasks?	Unexpected unplanned situations that arise as a result of faults, misconduct, errors, accidents, etc. are the most vulnerable to change. A large number of rail traffic participants, i.e., infrastructure managers, railway undertakings, entities in charge of maintenance, workshops and third parties, may influence changes in the railway system during its continuous operation.
3. Are the rest of the elements coordinated with respect to the goals/objectives and tasks?	All activities in the railway system are planned. Any influence of other parties must be agreed with the railway traffic operators or those maintaining the railway infrastructure, etc.
4. Where are the strengths and weaknesses in the interactions among elements?	Weaknesses occur at the interface of various devices in the form of physical and functional interfaces. Different types of devices can malfunction in the context of cooperation with other devices. The failure of one device may have unknown consequences. Also, interfaces between humans and devices can generate hazards due to a lack of training or knowledge of the device or subsystem being operated.
5. What are the sensitive points that repeatedly cause friction, problems, or conflicts?	Cooperation between different actors involved in real-time rail traffic may be a crucial issue. Different economic interests can also obscure the correct assessment of risks. For this purpose, certified rail traffic safety management systems are used which underpin the supervision of processes.

(Contd.)

No.	Parameters describing the socio-technical system for the ecosystem of railway transport companies
	3. Structures:
1. Are the existing structures useful for the attainment of the goals and tasks?	The existing safety management systems and management systems for the maintenance of rolling stock and railway infrastructure, as well as systems for authorizing equipment, systems for operation and systems supporting the work of train drivers and rail traffic controllers, are sufficient to ensure the core functions of operational readiness and rail traffic safety. Due to the processes of digital transformation, these processes are constantly being changed and new standards are being set in this area.
2. How do people adapt to or cope with these structures?	People are continuously trained in accordance with the requirements set out in legislation and training plans as part of safety management systems. People are the most important part of the railway system and are fundamental to achieving the objectives of the system and the companies operating within its borders with specific business models. They are supervised and treated as an essential part of the overall system.
3. Are the structures supportive of the needs and capabilities of the leadership (management) and the workforce?	The organizational structures of railway companies operating in the railway system are very rigorously subordinated to the processes – these positions are sometimes called services, because they have a direct impact on the safety of passengers, employees and third parties.
4. Do the structures encourage c o-operation and collaboration?	Organizational structures are geared towards cooperation and collaboration. The quality of the cooperation, including the exchange of information, determines the ability of the organization to perform its basic functions.
5. Do the structures facilitate a rational use of resources and technologies?	Yes, structures result from the geographical spread of roles in the railway system, and resources and technologies are the natural environment of the operator, who is obliged to know everything about the system's behavior and be able to react to normal situations, but also to operate in hazardous conditions, including when the system does not work as intended. The system must always be safe in accordance with the "Fail Safe" principle. That is how these systems are designed.
6. What major strengths/weaknesses result from the structures interacting with other elements in the system?	Strengths are based on many years of experience and the ways in which interfaces between devices and people are verified and validated. Weaknesses result from the ageing of technical parts and the slow process of replacing them with others with better performance and functions.

No.	**Parameters describing the socio-technical system for the ecosystem of railway transport companies**
	4. Leadership procedures:
1. Are the existing leadership procedures and processes focused on the specific goals/objectives and tasks?	As part of the legislation framework for safety and maintenance management system requirements, leadership must be demonstrated and ensure that all processes are supported by top management. To this end, appropriate communication and goal-setting procedures are implemented, as is building a positive organizational culture, including a culture of safety supported by management.
2. Do these procedures fit the organization's specific social and technical realities (technologies, other procedures, composition of the workforce, etc.)?	The procedures within the operation of companies in the railway system are specific and adapted to specific procedural realities. The level of alignment of procedures is assessed during audits and inspections by the National Safety Authority. Organizations have procedures in place to periodically verify, audit and monitor the correct application of appropriate operating procedures.
3. Are managers sufficiently trained in using the procedures (e.g., performance evaluation)?	Managers are regularly trained in accordance with the requirements of railway law. Top-level leadership delegates authority to lower levels and trains employees in the rules of conduct because of their responsibility for strategic and operational decisions.
4. What influences and changes from the outside are expected? What re-examination of the present procedures might these changes entail?	There are a number of risks associated with political, social, economic, and military crises and state governance priorities. Rail is a strategic system that reflects the state of the economy and is the first to detect signs of either crisis or growth.
5. What are the most significant strengths and weaknesses in the leadership procedures and their application?	Poor leadership may be related to the lack of understanding by some managers of the specificity of the operation of railway companies, the poor competence of staff or political influence in strategic railway companies.

(Contd.)

No.	Parameters describing the socio-technical system for the ecosystem of railway transport companies
5. People:	
1. What is the basic attitude of various employee groups towards the organization?	The attitude is subordinate, especially for posts directly related to railway traffic safety. Other positions are prominent in the context of the conditions related to the strategic role of these companies in the economy and the impact of work on organizational efficiency, the reliability of equipment operation and process safety.
2. What are the effects of these attitudes on performance and teamwork?	It is crucial to the levels of cooperation and performance in the operation of the railway system and rail traffic safety. Teams are trained together and management pays attention to the high level of awareness among the crew regarding the accountability of results and responsibility for the tasks performed.
3. How do managers and their subordinates adapt to the existing structures?	Managers understand that railway companies fulfill the requirements of a high reliability organization, so they use commonly known schemes of com-petence improvement based on the principles of benchmarking.
4. Do the employees feel that the structures, technologies and procedures serve as an incentive to full involvement (motivational effectiveness)?	Yes, motivational effectiveness is crucial and so are the responsibility structures, including the principles of employee replacements to maintain the continuity of the staffing of key positions in operational processes and management.
5. Where are the biggest obstacles, problem areas and sources of conflict?	There are procedures in place to deal with obstacles to achieving the objectives set. The procedures also include rules for resolving conflicts, in particular where they may have an impact on rail traffic safety.
6. Technologies:	
1. How do changes in procedures and support systems affect people and structures in the organization?	Changes in procedures are assessed in terms of their impact on rail traffic safety and organizational efficiency. Moreover, changes are assessed in terms of their impact on maintaining the competence of staff to carry out current tasks. Changes are also assessed in terms of their impact on the strategic functioning of the organization.

No.	Parameters describing the socio-technical system for the ecosystem of railway transport companies
2. Do changes produce resistance?	As a rule, changes always generate resistance. In these organizations, changes are assessed in terms of the system and employees understand that changes must be made to maintain business continuity, the operational readiness of the railway system and the effectiveness and efficiency of companies.
3. When technological changes occur, are the structures and processes adapted or modified accordingly?	In rail transport, technological progress is crucial. However, innovation is implemented slowly. Technological changes usually cause significant changes in work structures, organizational structures and processes in this sector. For example, the implementation of ERTMS and GSM-R train control and supervision systems or other such standards require significant changes in the level of training, but also organizational changes at strategic and operational levels. They also require cooperation between undertakings and rail infrastructure managers at strategic, tactical and operational levels.
4. Is there a timely attempt to prepare people for technological and organizational changes, e.g., through personnel management, training or participatory decision making?	In the case of strategic projects, technological changes are implemented systemically and the crew is prepared in many respects to exercise diligence in the implementation of changes. Each change requires the extensive preparation of personnel and adaptation of operational processes for proper service. Appropriate procedures apply that are strictly followed and their degree of accuracy is assessed during periodic audits of management systems.
5. What and where are the most important strengths and weaknesses between technology (including processes and procedures) and the other elements of the system?	The weakest side is human awareness of the place and role of technology in the process. Therefore, these companies focus special attention on staff training and building awareness and knowledge as well as the technical culture of the organization. Human error can always occur, so one should do everything to reduce the likelihood of it occurring. There are also psychophysical tests and periodic assessments of employees in terms of perception and teaching behavior in critical situations—non-standard, for example, on locomotive simulators or the workstations of a rail traffic controller or employees performing rolling stock maneuvers.

Source: Own study based on Appelbaum (1997, pp. 452–463).

Table 7.4 Parameters Describing the Socio-technical System for the Ecosystem of Companies in the Water and Sewage Sector

No.	Parameters describing the socio-technical system for the ecosystem of companies in the water and sewage sector
	1. Environment:
1. What are the major factors in the organization's environment that influences the work?	The main factors that affect the functioning of the organization is maintaining the efficiency of the water supply and sewage network, paying special attention to the mass of recipients who use this resource. Supervision of these processes is crucial and work processes are subordinated to it.
2. What characteristic environmental problems arise?	In the water sector, companies have a direct impact on the environment by directly taking water resources, converting them into drinking water and collecting wastewater that they treat in appropriate wastewater treatment plants. The process is crucial for the protection of the environment and is bound by many legal provisions in this area.
3. In which elements of the system are the interactions with the environment most visible?	The most visible interactions with the environment concern sewage treatment plants, the tightness of pipelines, the efficiency of the water sewage system, monitoring of sewage, monitoring of waste, the impact on air pollution and the introduction of sewage into the soil, which is prohibited. Responsibility in this area is very high and large financial outlays are required to ensure the continuity of operation of water and sewage systems compliant with environmental regulations.
4. What is the organization's greatest strength *vis-à-vis* the environment (competitive advantage, labor market, productivity, etc.)?	Water and sewage companies have a privileged position on the local market, as they meet the requirements of a natural monopoly. They are very often key employers and have a significant impact on the social environment. Water quality and efficiencies in wastewater collection are fundamental to ensuring the quality of life of citizens.
	2. Goals/tasks:
1. What factors within the system most influence goal attainment (positively/negatively)?	Crucial factors that affect the attainment of strategic goals are the size of the sewage network, sources of income, the possibility of obtaining subsidies, support for municipal and city management, the wealth of citizens and the ability to accept the high price of water, the quality of management of the company and the support of city and commune boards.

No.	Parameters describing the socio-technical system for the ecosystem of companies in the water and sewage sector
2. Which elements of the system are most affected by changes in goals and tasks?	The key elements that affect the changes related to the achievement of the goals are the efficiency of the water sewage system depending on its technical condition and the quality of repair services. The key is also good organization of work, understood as unity in setting priorities to ensure the continuity of the water and sewage system.
3. Are the rest of the elements coordinated with respect to the goals/objectives and asks?	It is crucial to monitor operational processes, including the use of key performance indicators, for which critical values should not be exceeded. Very often, the model of a strategic balanced scorecard focused on technological and organizational aspects is used in water supply companies.
4. Where are the strengths and weaknesses in the interactions among elements?	A strong point is the direct connection of employee structures to the water supply and sewage network, as the purpose of such an organization is to ensure its efficiency for water supply and sewage collection. Auxiliary processes such as water quality tests in accredited laboratories are important and serve to support new processes of serving citizens of specific areas of operation of these companies.
5. What are the sensitive points that repeatedly cause friction, problems, or conflicts?	The main problems are the relationships between water users and sewage suppliers and the company, which is very often supervised by the city or commune authorities. The relationships between these three stakeholders require informed and responsible decisions.
3. Structures:	
1. Are the existing structures useful for the attainment of the goals and tasks?	In water and sewage companies, the structures of the manager look different. As a rule, they consist of a strategic division that is responsible for investments and an operational division that is responsible for ongoing business continuity assurance in terms of water supply and sewage collection. Simple management through a system of simple rules is very often used here.
2. How do people adapt to or cope with these structures?	People adapt to the implemented structures because the core function of the socio-technical system is to ensure the continuity of the operation of the water and sewage system and everything is subordinated to this purpose.

(Contd.)

No.	Parameters describing the socio-technical system for the ecosystem of companies in the water and sewage sector
3. Are the structures supportive of the needs and capabilities of the leadership (management) and the workforce?	Managers in water and sewage companies should show leadership, as they play a service role towards citizens and play a very important role in their quality of life. Leadership in the area of responsibility for delivered values should be visible in the community of recipients of these services.
4. Do the structures encourage co-operation and collaboration?	Structures in such companies are very simple, although given that they are strategic for the functioning of municipalities, they should be supported by large groups of experts who provide knowledge that is directly used in operational processes.
5. Do the structures facilitate a rational use of resources and technologies?	Structures are not always effective. In order to be effective, they must be based on good assumptions. A theoretical error that is based on making bad assumptions should not be made. This can lead to a practical error. Operational intellectual support is essential for this type of organization.
6. What major strengths/weaknesses result from the structures interacting with other elements in the system?	Strengths in the interaction of organizational structures with elements of the socio-technical system are the ability of the system to perform core tasks. The human and organizational factors displaced by well-accepted technologies, both from the sphere of handling water and sewage services and waste treatment within the framework of environmental protection regulations, require leadership and organizational support.
4. Leadership Procedures:	
1. Are the existing leadership procedures and processes focused on the specific goals/objectives and tasks?	Companies have the most frequently implemented quality management systems, which are used to ensure that all activities are carried out in accordance with the plan and adopted standards and norms. All types of risks of error leading to a dysfunction of the water supply system that could lead to failure are monitored.
2. Do these procedures fit the organization's specific social and technical realities (technologies, other procedures, composition of the workforce, etc.)?	The procedures are usually tailored to the organizational structures and the staffing of workplaces. Technologies need to be overseen through procedures developed as part of a process approach. The workload should be relevant to the level of complexity of processes, the scope of technologies used, the spatial arrangement of water supply points and sewage collection.

No.	Parameters describing the socio-technical system for the ecosystem of companies in the water and sewage sector
3. Are managers sufficiently trained in using the procedures (e.g., performance evaluation)?	Managers are usually well trained in the implementation of supervisory activities related to the process of handling the water supply process and sewage collection. They have strategic plans for the implementation of infrastructure renewal and network modernization investments. They evaluate the results periodically for the purposes of reporting by the supervisory authorities.
4. What influences and changes from the outside are expected? What re-examination of the present procedures might these changes entail?	The changes in water and sewage companies are currently focused on cost optimization due to the increased cost of raw materials and utilities. The maintenance of a reasonable price is influenced by the search for savings in processes using digital innovations, including in the field of the circular economy. Digital transformation is conducive to improving performance monitoring. These changes are necessary to maintain the continuity of the system, to maintain the viability of companies and to implement innovations based on technological progress.
5. What are the most significant strengths and weaknesses in the leadership procedures and their application?	In water and sewage companies, a so-called concept of limited leadership appears, which results directly from the supervision of municipal authorities, the influence of external stakeholders on decision-making processes and the role of service recipients in the acceptance or non-acceptance of the price for these utilities. Social dialogue is required to eliminate conflict situations.
5. People	
1. What is the basic attitude of various employee groups towards the organization?	Employees working in companies of this type have stable employment because they work in companies that have a natural monopoly. Stable employment is key to maintaining staff competence and process efficiency.
2. What are the effects of these attitudes on performance and teamwork?	The efficiency of work depends on the rules of work standardization and accepted rates for the work performed. In some companies it is implemented correctly and does not arouse controversy; in others, on the other hand, the development of fair procedures for remunerating employees and building project-oriented teams is required.

(Contd.)

No.	Parameters describing the socio-technical system for the ecosystem of companies in the water and sewage sector
3. How do managers and their subordinates adapt to the existing structures?	Managers and subordinates in such organizations quickly adapt to changes. Some organizations have trade unions that support workers in making decisions. Corporate Social Responsibility (CSR) principles, including ESG procedures, are applied in many such organizations.
4. Do the employees feel that the structures, technologies and procedures serve as an incentive to full involvement (motivational effectiveness)?	The observation of companies demonstrated that organizational structures match incentive systems, and incentive tools fulfill their roles. Managers try to motivate employees to work and make this motivation dependent on the efficiency of the sewage network.
5. Where are the biggest obstacles, problem areas and sources of conflict?	Most of the sources of conflict are in the area of remuneration and labor regulation. Labor systems should be fair and not generate conflicts. Managers take care to maintain clear procedures for resolving conflicts and clarifying disputes.
6. Technologies:	
1. How do changes in procedures and support systems affect people and structures in the organization?	Changes in procedures related to the entire socio-technical system on the basis of which water and sewage companies are built influence and pressure managers on the implementation of new technologies. It is quite common for managers to search for new technologies in the field of renewable energy sources, environmental protection, digital transformation and the Industry 4.0 concept. This is crucial to ensuring the development of this type of organization in terms of the efficiency and effectiveness of management.
2. Do changes produce resistance?	They always generate resistance, but good managers can convince the crew that changes are necessary, especially in the area of technology. Technological progress and dynamics of changes, as well as the need to optimize costs, require the implementation of digital transformation and lean management solutions.
3. When technological changes occur, are the structures and processes adapted or modified accordingly	Technological changes have a crucial impact on the processes of adapting procedures and structures. Modern digital technologies improve communication systems, implement controlling procedures and improve operational efficiency. The adaptation of new regulations in the area of environmental protection and orientation towards business models affecting positive climate change is a challenge for water and sewage companies.

No.	Parameters describing the socio-technical system for the ecosystem of companies in the water and sewage sector
4. Is there a timely attempt to prepare people for technological and organizational changes, e.g., through personnel management, training or participatory decision making?	Training campaigns are carried out to prepare employees for changes resulting from technological progress, digital transformation, new business models operating in the water and sewage sector, new business models in the field of renewable energy and modern technologies conducive to cost optimization and error neutralization. Employees are trained in risk management in order to prevent the occurrence of disaster, failures and dysfunctions in the functioning of the system.
5. What and where are the most important strengths and weaknesses between technology (including processes and procedures) and the other elements of the system?	A strength in the implementation of new technologies is their impact on cost optimization. A weakness, however, is the lack of funds due to operational problems. The European Union's support for the improvement of processes related to environmental protection and climate change can strongly neutralize the risk of investment.

Source: Own study based on Appelbaum (1997, pp. 452–463).

the creation of digital platforms, business models and point-of-use technological solutions creating new innovations on the energy market. The base canvas for describing digital business models/digital platforms as socio-technical systems are shown in Figure 7.1.

Based on standardized questions related to the assessment of socio-technical systems, the characteristics of the ecosystem of energy sector companies were developed. To this end, the questions developed by Appelbaum (Appelbaum 1997, pp. 452–463) were used what shows Table 7.5.

7.4 THE AUTHOR'S ORIGINAL CANVAS OF THE DESCRIPTION OF THE DIGITAL BUSINESS MODEL BASED ON THE ASSUMPTIONS OF SOCIO-TECHNICAL SYSTEMS

Two core layers built into the next two sub-layers, namely the theoretical layer and the practical layer, have been taken into account for a comprehensive baseline description of issues related to the description of digital business models/platforms as socio-technical systems. The first theoretical layer consists of an indication of the areas that fill the scope of the concept of socio-technical systems and consists of six components, namely People, Technology, Infrastructure, Process, Goals and Culture. The second theoretical layer is the scientific achievements in the field of complex systems because the digital ecosystems of the business model are complex and fit into this theory. It consists of areas such as Networks, Evolution and Adoption, Pattern Formation, System Theory, Nonlinear Dynamics, Game Theory and Collective Behavior. The practical layer was built from two sublayers, i.e., a layer of description of traditional business models built on the assumptions of the canvas by Osterwalder and Pigneur (Osterwalder and Pigneur 2010) and a sublayer relevant to the digital economy by Blaschke and his team (Blaschke et al. 2017, pp. 121–136).

The layer, in terms of socio-technical systems and in terms of complex systems, aims to explain how digital business models function in the complex business ecosystem. They cannot be so easily described without understanding the theoretical assumptions and taking them into account in the process of logical reasoning. Only with a full understanding of the theoretical assumptions and the identification of all the components describing the theoretical layers can one proceed to the identification of practical aspects that are based on two popular and widely accepted methods of business model notation. The first is the practical sublayer, which consists of nine elements defined by

Osterwalder and Pigneur. It is complemented by six elements proposed by Blaschke and his team (Blaschke et al. 2017) and directly refers to the specificity of the activities of digital platforms and technological solutions based on digital transformation. Such a complex arrangement of links is a proposal that can explain to a large extent how to identify, treat, describe and explain today's digital business models based on platformization technology.

7.5 DEVELOPMENT OF THE AUTHOR'S ORIGINAL METHODOLOGY FOR DESIGNING DIGITAL BUSINESS MODELS—DIGITAL PLATFORMS BASED ON THE ASSUMPTIONS OF THE CONCEPT OF SOCIO-TECHNICAL SYSTEMS

Complex systems, and in fact ecosystems of ecosystems, are based on holism, speaking of the behavior of the system which is fully determined by its components. However, in order to obtain an object in the design process fully composed of numerous functions, it is necessary to separate such components that are crucial to achieving the planned functions and performance of the designed business models. In turn, the approach of disassembling systems into parts and establishing relationships between them, including interactions, in order for them to work together as a whole is called reductionism. Both holism and reductionism are necessary to carry out the process of designing digital platforms operating in a specific business ecosystem. The design of digital platforms should be based on agile project management methodologies. Due to the fact that these are socio-technical systems, a systemic approach to the design of such systems can be applied, namely the V-model. Figure 7.2 is the graphical presentation of the author's original methodology for designing digital business models—digital platforms based on the assumptions of the concept of socio-technical systems.

The methodology for designing digital business models—digital platforms—was developed with three core assumptions in mind. The first assumption was to use the V-model from the rail transport sector recommended by the CENELEC (European Committee for Electrotechnical Standardization) series of technical standards to describe the methodology. The second assumption that underlies the methodology was the use of agile project management methodology, the scope of which was integrated into the structure of the V-model. The third element included in the methodology process of designing digital business models is the use of socio-technical system criteria, namely environmental, goals/tasks, structures, leadership, people

Table 7.5 Parameters Describing the Socio-technical System for the Ecosystem of the Energy Sector

No.	Parameters describing the socio-technical system for the ecosystem of the energy sector
	1. Environment:
1. What are the major factors in the organization's environment that influences the work?	The environment of the functioning of companies in the energy sector is based on functioning in a technological and network environment, where the core business model is the business model of the distribution system operator supplying energy to an institutional or individual customer. Processes are associated with a strong focus on occupational health and safety requirements, as the working environment is dangerous and poses a large number of risks to employees.
2. What characteristic environmental problems arise?	The energy ecosystem has a crucial impact on the natural environment and is shaped by a set of technical installations for which it is necessary to ensure safe operation and maintenance conditions. Key problems are based on the replacement of old traditional electricity installations with new decentralized solutions based on renewable energy sources.
3. In which elements of the system are the interactions with the environment most visible?	An electricity grid is a set of interrelated elements that interact with each other in a specific energy environment. Surveillance by operators in a distributed geographical system is essential and requires good work organization. All process sites must be monitored and evaluated. All the links in the system form a comprehensive picture of the efficiency of the system.
4. What is the organization's greatest strength *vis-à-vis* the environment (competitive advantage, labor market, productivity, etc.)?	Energy companies are, as a rule, natural monopolies as regards traditional energy. The situation becomes complicated when distributed sources of energy enter the game in the form of self-sufficient clusters. The decentralization of generation of energy sources with the participation of prosumers completely changes the picture of competitiveness in this sector. Revolutionary changes in the energy market forced by the need to implement a climate-changing business model determine the emergence of new forms of energy generation based on new organizational formations. In this reality, the productivity, efficiency and effectiveness of the proposed solutions change.
5. What is the organization's most obvious weakness?	Weaknesses of the organization result directly from dynamic changes in legislation, with an emphasis on changes in the field of power engineering, especially due to ecological criteria and high energy prices. The social environment requires energy prices to fall and the efficiency of the electricity system to increase. Business models need to be more customer-friendly and cheaper.

No.	Parameters describing the socio-technical system for the ecosystem of the energy sector
2. Goals/tasks:	
1. What factors within the system most influence goal attainment (positively/negatively)?	The impact of technological innovations changing the rules of the game in the sector affects the development of the sector and its companies. The system is changing due to the emergence of new forms of energy generation and delivery to the end consumer. At the same time, a number of innovative solutions based on energy production on local markets occur. The positive aspect of traditional generation and distribution of energy is that there is no energy storage technology yet, which gives an advantage to large plants.
2. Which elements of the system are most affected by changes in goals and tasks?	Power systems that are based on traditional energy are the most vulnerable to changes and will certainly in the near future be completely replaced by renewable energy sources. Distributed energy creates new conditions and opportunities for achieving energy goals.
3. Are the rest of the elements coordinated with respect to the goals/objectives and tasks?	The power system is interdependent, i.e., power generators, a distribution system operator, a high-voltage grid operator, and a seller must work together to achieve synergy between business objectives, social impact and expectations from the sphere of environmental benefits.
4. Where are the strengths and weaknesses in the interactions among elements?	This is a disadvantage and these dependencies are too strong in relation to large enterprises, which inhibits the development of small power plants and local distribution networks. Legal and technological barriers hinder the development of distributed energy. Such complex socio-technical systems require further analysis, development and the construction of new business models.
5. What are the sensitive points that repeatedly cause friction, problems, or conflicts?	The crucial problems are based on communication difficulties and conflicts of interest in achieving business objectives, including the economic benefits resulting directly from the sharing of the distribution network and mutual needs in terms of target customer service. This is directly due to the revolutionary nature of the changes in this sector and the change in the field of power between the various network players and actors.
3. Structures:	
1. Are the existing structures useful for the attainment of the goals and tasks?	As regards the power system, different structures with specific objectives and tasks occur. The complexity of these structures is highly dependent on the business model adopted for the generation and supply of electricity to the end consumer.

(Contd.)

No.	Parameters describing the socio-technical system for the ecosystem of the energy sector
2. How do people adapt to or cope with these structures?	Adaptation to new structures is based on the assumptions of Schumpeter's creative destruction, where new solutions displace the old, and the old ones are still doing well. In this way, people must adapt to solutions that require, on the one hand, acceptance of innovation and progress, and on the other hand, patience to transform the power system into an innovative, beneficial model.
3. Are the structures supportive of the needs and capabilities of the leadership (management) and the workforce?	Modern management models include structures supporting the need to improve efficiency based on leadership and unambiguous indication of the direction of company development. Leadership supports work processes and the creation of innovative solutions in the technical and organizational sphere.
4. Do the structures encourage co-operation and collaboration?	The structures of companies operating in the energy sector, on the one hand, are focused on cooperation in the contractual model and, on the other hand, defend their interests. A competitive aspect and the effect of cooperation occur at the same time. Typical network coopetition occurs.
5. Do the structures facilitate a rational use of resources and technologies?	Energy company structures are used to rationally use resources and technologies to ensure business continuity and implement progressive solutions.
6. What major strengths/weaknesses result from the structures interacting with other elements in the system?	The strengths are the well-established position of the energy company in the monopoly market, albeit it is threatened by the creation of bottom-up initiatives to create self-sufficient energy clusters. Thus, structures and interactions change, creating new assumptions for prosumer power engineering.
4. Leadership procedures:	
1. Are the existing leadership procedures and processes focused on the specific goals/objectives and tasks?	Leadership is based on the intellectual building of authority through knowledge and its grounding. Goals are only achieved if they are supported by well-established knowledge. In this sector, intellectual power that determines the attitudes of modern leaders can be referred to.
2. Do these procedures fit the organization's specific social and technical realities (technologies, other procedures, composition of the workforce, etc.)?	The procedures are adapted to contemporary realities in social and technical terms, taking into account modern technologies underlying the creation of new working conditions. Technologies are described in operational procedures as part of a process approach.

No.	Parameters describing the socio-technical system for the ecosystem of the energy sector
3. Are managers sufficiently trained in using the procedures (e.g., performance evaluation)?	Managers are well trained in the use of procedures and the evaluation of results through the use of modern monitoring methods taking into account digital transformation solutions. Software dedicated to individual processes is centralized and systemically used.
4. What influences and changes from the outside are expected? What re-examination of the present procedures might these changes entail?	Changes in procedures are influenced by technological progress, customer requirements, legal regulations and a positive impact on changes towards building modern energy based on renewable energy sources. The parallel character of these processes creates the need for dynamic changes in the methods of value delivery and customer service. The needs in terms of technology use and efficiency are also changing.
5. What are the most significant strengths and weaknesses in the leadership procedures and their application?	The key problems arising from leadership are fusion syndrome with old solutions and difficulties in implementing new ones. A strength is the search for new innovative solutions that compete with old structures. At the same time, the socio-technical system is changing significantly, forced by the need to optimize costs and create a modern energy industry that sells to business models that have a positive impact on climate change.
	5. People:
1. What is the basic attitude of various employee groups towards the organization?	In modernly managed companies, there is a full understanding for various groups of internal and external stakeholders who expect specific organizational changes that are conducive to their interests. Balancing these goals is a core task for managers.
2. What are the effects of these attitudes on performance and teamwork?	Modern management methods based on leadership, especially positive leadership, ensure that employees achieve the expected results and teams work in synergy. In power companies, the strategic balanced scorecard model is very often implemented.
3. How do managers and their subordinates adapt to the existing structures?	Managers try to ensure the coordination of current structures in the context of operating conditions based on ensuring the continuity of the power system, while maintaining the criteria of health and safety at work and employee motivation.
4. Do the employees feel that the structures, technologies and procedures serve as an incentive to full involvement (motivational effectiveness)?	In the energy sector, solutions based on project orientation are currently being implemented in the environment of many projects, which translates into improved work efficiency and changes in procedures. The position of many projects allows for the creation of a portfolio in projects, programs and solutions to improve the effectiveness of project management. In this formula, the effectiveness and efficiency of the organization is enhanced.

(Contd.)

No.	Parameters describing the socio-technical system for the ecosystem of the energy sector
5. Where are the biggest obstacles, problem areas and sources of conflict?	As in any organization, key obstacles result directly from character traits that are invariable, as evidenced by the study of psychological cybernetics; therefore, managers use a number of solutions aimed at avoiding conflicts and points of contention.
6. Technologies:	
1. How do changes in procedures and support systems affect people and structures in the organization?	Changes in procedures significantly affect the effectiveness of employees, especially since they often concern changes in organizational structures. Any change causes anxiety among employees; therefore, discussions are required that clarify and build knowledge and awareness of the need for change in the context of challenges faced by the energy sector for the survival of companies and ensuring their business continuity.
2. Do changes produce resistance?	Every change in the organization creates resistance. How the change is explained, implemented and assessed depends on leaders. Change management is a process that requires convincing staff and demonstrating the benefits of implementing change.
3. When technological changes occur, are the structures and processes adapted or modified accordingly?	Along with the processes of implementing new technologies in the field of energy, structures and processes are changed accordingly, and procedures for adapting these solutions to the business management practice are modified. Each technological change is translated into a procedural solution, which serves to better adapt and explain these changes.
4. Is there a timely attempt to prepare people for technological and organizational changes, e.g., through personnel management, training or participatory decision making?	As part of the project orientation, participatory systems are used to involve project team members in the decision-making process at various levels of management, taking into account the strategic development needs of companies. Training and integration meetings are held and brochures and other information materials are published to explain the changes. Managers make use of lower-level employees in the decision-making process.
5. What and where are the most important strengths and weaknesses between technology (including processes and procedures) and the other elements of the system?	The weakness of the technologies used is that they are not fully known and significantly affect the entire business model of the energy company. The change of technology is not only a point of change, but significantly affects the entire power system, transforming current procedures into new formulas of action. This is of fundamental importance to understanding changes and the transformation of traditional energy to renewable energy, including decentralized energy. A strength is the strong imperative to build energy competitiveness on global markets.

Source: Own study based on Appelbaum (1997, pp. 452–463).

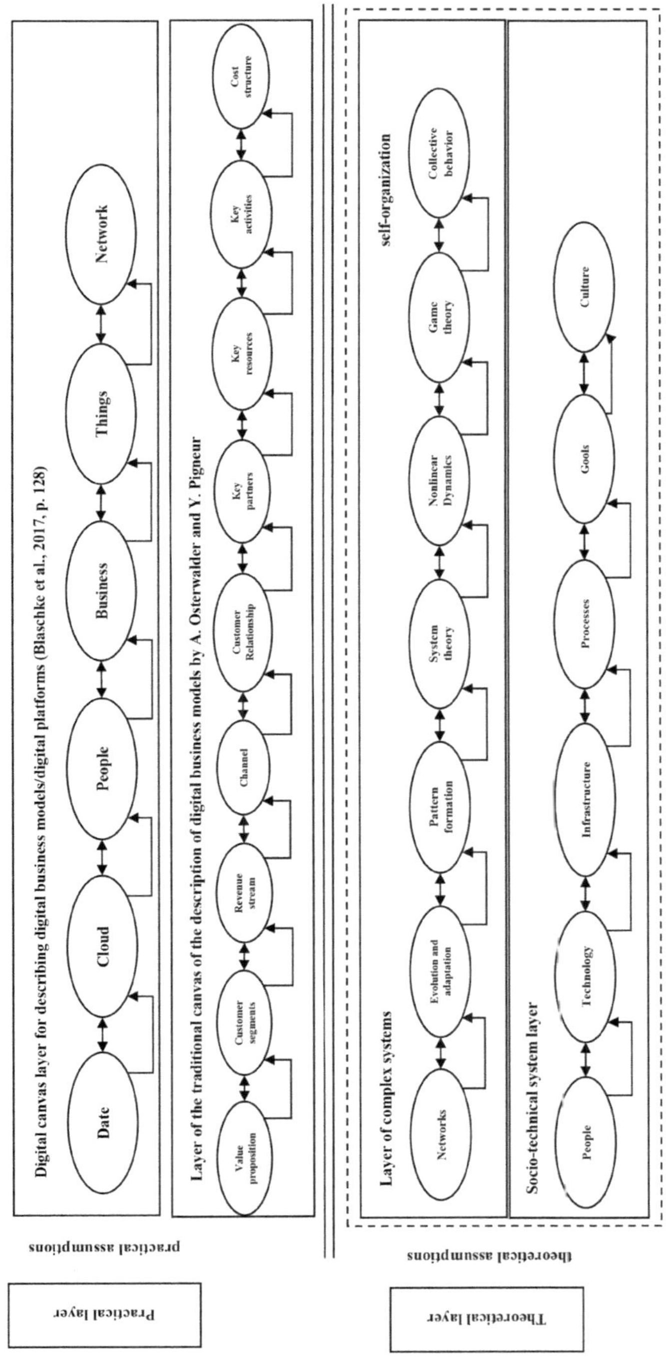

Figure 7.1 The base canvas for describing digital business models/digital platforms as socio-technical systems.
Source: Own study.

and technologies, which were considered criteria for the operational requirements of the designed solution. The condition for achieving the expected effect in accordance with the customer's requirements, taking into account the popular attractive social idea, is the implementation of periodic sprints—intensified activities related to the implementation of the project and verification of the effects of this sprint with regard to customer expectations. As the project is implemented, there are increases in time, product, costs and labor. All these activities are carried out according to the to-do ranking—product backlog. A product backlog is a structured list of works planned for the development team, which results from the product action plan and its requirements. All activities follow the logic of project implementation according to the V-model, according to which the project starts with a popular and attractive idea, and then this idea is translated into the next stages of project implementation, which include:

1. Defining the scope of the digital platform project, selecting the technology and developing a project plan.
2. Assessing project risk—development of a hazard and risk register.
3. Specifying customer requirements and other requirements including legal requirements.
4. Developing the project and defining technical and organizational assumptions for the implementation of solutions.
5. Producing the digital platform, selecting technologies, conducting technology fusions with other owners of other required technologies, if justified, and developing a prototype of the digital platform.
6. Installing a digital platform prototype in a testing environment.
7. Solution verification and validation – V&V process.
8. Acceptance of the proposed socio-technical system of the digital platform.
9. Acceptance of the system.
10. Operation and maintenance.

According to the V-model, the final stage is the decommissioning and disposal of solution after its life cycle.

A methodology of designing digital business models so designed fulfills the requirements of holism and reductionism at the same time. Such opposing theories are possible for simultaneous implementation only due to the use of an iterative approach to search for the optimal solution using sprints derived from agile project management methodologies. Through sprints, it is possible to simultaneously reduce and add elements

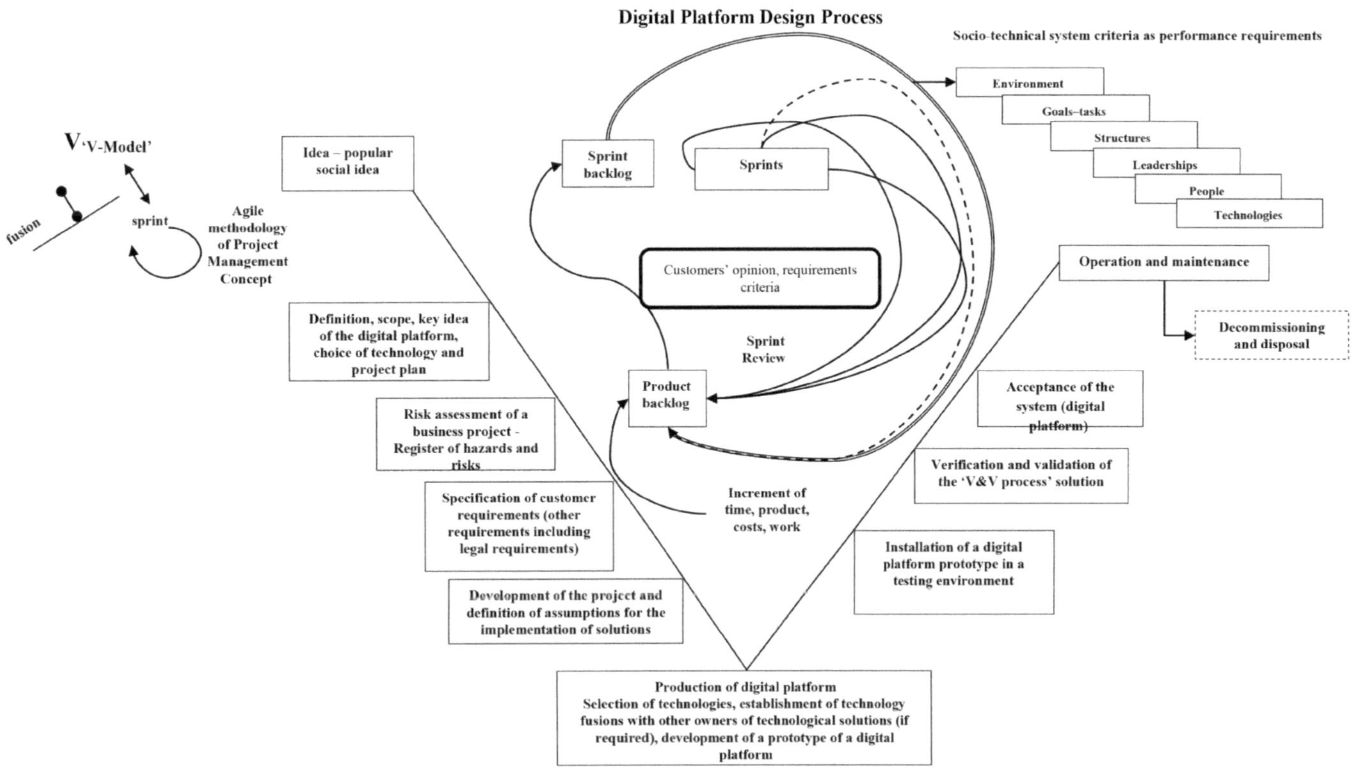

Figure 7.2 Methodology for designing digital business models—digital platforms based on the assumptions of socio-technical systems. (*Source:* Own study based on EN 50126-1 RAMS 2021, p. 54).

in order to obtain an optimal system operating in a specific business ecosystem that is part of other ecosystems within the framework of complexity theory. Conducting such a structured process of designing digital platforms in accordance with the proposed methodology should ensure not only that the customer's requirements are met and an attractive product is launched on the market, but also that it is fully verified and validated in terms of technical safety and cybersecurity.

7.6 DEVELOPMENT OF A SET OF MANAGEMENT AND TECHNICAL-OPERATIONAL FACTORS THAT HAVE THE GREATEST IMPACT ON THE EFFECTIVENESS OF DIGITAL BUSINESS MODELS

Digital business models can be viewed from two perspectives. The first perspective concerns the issue of building a component configuration of business models based on the logic of generating value for customers and is based on the assumption that Big Data sets concerning mass recipients—digital platform users—contribute to the creation of value. The second perspective, on the other hand, refers to the concept of socio-technical systems, where the issue of interaction between people is possible only through specific digital technologies. These two perspectives demonstrate the need to combine the design engineering of digital platforms with softer elements, which include the aspects of socio-technical systems. A balanced focus on human and technology perspectives allows for the design of solutions that, on the one hand, are attractive ideas acceptable to large human communities, but, on the other hand, are easy to use, widely available and socially acceptable. In this context, a need emerges to define crucial management and technical–operational factors that will indicate the limits of the use of socio-technical systems in the design and development of digital business models based on digital platforms.

Rhodes and Ross developed five strategies which use a taxonomy of five aspects influencing the creation of innovations in social engineering systems (Rhodes and Ross 2010). It covers the following thematic areas:

1. **Structural factors**—related to the form of system components and their interrelationships.
2. **Behavioral factors**—related to performance, operations and responses to underlying stimuli for the current "state of practice" of systems architecture and design and model-based systems engineering.

3. **Contextual factors**—related to the circumstances in which the system exists.
4. **Time factors**—related to the dimensions/properties of the systems over time.
5. **Perceptual factors**—related to stakeholder preferences.

Based on the abovementioned factors, they were interpreted for the purpose of designing digital platforms for which it was assumed that they operate in the sharing economy, the circular economy or Big Data systems. They can also be based on other trends resulting from the assumptions of the digital economy. They should be based on the assumptions of socio-technical systems and be subject to the following:

1. **Conceptualization:** Recognizing and verbalizing the research problem concerning the relationship between selected phenomena, formulating a preliminary hypothesis and (possibly) showing the circumstances of the problem. Preliminary identification and assessment of the problem and justification for addressing it—usually this is due to the discovery of some irregularities.
2. **Specification:** Positioning the problem and hypothesis in a theoretical context, defining the assumptions and formulating detailed research hypotheses in terms of relationships between the separated variables.
3. **Operationalization:** Determination of criteria of empirical applicability and quantitative recognition for non-observable terms occurring in detailed hypotheses. Planning (or selecting an appropriate model) of diagnostic procedure and selection of appropriate (i.e., relevant to hypotheses) diagnostic methods.
4. **Comparison:** Selection of models of collecting and organizing experience compared with the content of empirical implications of hypotheses and models of data analysis. Application of selected research and statistical models. Collection of data by means of selected methods and analysis of results in terms of their compliance with the diagnostic hypotheses.
5. **Application:** Formulation of conclusions on the veracity of the tested hypotheses and the indication of the acceptable reference range of the obtained research results to:
    1. assumptions and research hypotheses;
    2. the possibility of formulating further problems, possibly proposing theoretical and methodological postulates and revising the existing paradigm;
    3. taking practical actions based on them (Spendel 2005, p. 33). Selected factors of socio-technical systems aggregated for the

management of technical and operational aspects with the greatest impact on the efficiency of digital business models are shown in Table 7.6.

The proposed system for defining requirements and conditions for the design of digital platforms, taking into account social technical factors, demonstrates its complexity and difficulty. In the process of designing technical social systems, it is necessary to take into account not only issues related to the configuration of the designed business model and its operating technology, but also the factors of future system operation taking into account the needs of users and stakeholders, the issue of applying technology to the functionality of the system and the conditions of cooperation of the digital platform with other platforms. It is also important to take into account the factors of ergonomics of work in relation to the planned tasks of the system and the social acceptance of the designed solution.

7.7 DISCUSSION OF RESEARCH RESULTS—CONCLUSIONS FOR THEORY, PRACTICE, FUTURE DIRECTIONS OF RESEARCH AND LIMITATIONS OF THE RESEARCH PROCESS, THEORETICAL IMPLICATIONS, PRACTICAL IMPLICATIONS, FUTURE, POTENTIALLY INTERESTING DIRECTIONS OF RESEARCH, LIMITATIONS OF THE RESEARCH

The main aim of the scientific discussion was to demonstrate the determinants of shaping digital business models and complex technological companies with the potential to develop the use of multi-sided digital platforms as part of technological development based on the assumptions of the theory of socio-technical systems. The analyses, systematic and critical literature review and conceptual work created as a result of literature studies and conclusions, as well as the author's practical expertise in the field of socio-technical systems design, provided an opportunity to draw a number of conclusions, offer recommendations and propose several applicable solutions. As a result of the work, the scientific achievements in this field were significantly expanded, which allows for the use of these concepts in new conditions based on digital platforms and applied technologies that fit into the assumptions of the network paradigm and the theory of complexity of systems. As part of the scientific goal, it is crucial to determine the scientific problem and identify the cognitive gap. This cognitive gap was an identifiable set of factors responsible for the effective creation and management of the development of companies, which are based on the assumptions of business models built on the use of digital transformation, where technological factors and business platformization shape the business environment based on the assumptions of the functioning of socio-technical systems. In doing so, the book demonstrated a number of different approaches and concepts in the field of socio-technical systems, the taxonomy of digital platforms, complexity theory and many other approaches including the description of selected business models focused on social and ecological aspects. The cognitive scope so defined gave rise to the emergence of a new concept of designing digital business models suitable not only for innovative solutions in the sphere of information and communication and digital platforms supporting innovative business models, but also in the field of traditional forms of doing business in the railway, energy or water supply sectors, where technical infrastructure determines the conditions for value delivery to recipients, and a community is built alongside. In particular, these business ecosystems meet the definition of socio-technical systems in theoretical and practical terms. The author of the monograph proved that engineering, technological solutions and digital transformation break the current paradigms of strategic and operational management and should be explained again by means of different assumptions and principles of describing modern business, including its limitations. At the same time, there is a need to combine social and technical aspects in parallel to explain phenomena in the area of new theories and build utilitarian solutions based on the formation of complex systems which underlie the environment based on the concept of a system of systems and even ecosystems. As indicated in research assumptions, omitting engineering aspects and relying on only the so far known and well described paradigms of management and quality science without interdisciplinary technological support from other disciplines such as safety sciences, management psychology, and sociology is a mistake and leads nowhere. What has been recognized in the management and quality sciences as paradigms in the era of widespread digital transformation becomes archaic and useless for building new theories and creating application solutions useful to management practitioners, which is vitally important to build a high position of science in economic and social life. In this way, space for reconfiguring current paradigms and building new approaches and concepts based on socio-technical systems which are considered in terms of the functioning of the digital economy and new ideas resulting from it emerges. In the area of research and the scientific discussion, a systematic literature review on the concepts of the socio-technical system in the business model was conducted, which was the basis on which a number of original approaches to the discussed topic were developed. Several case studies were described, both in the field of "purely" digital business, e.g., multi-sided digital platforms, and case

Table 7.6 Selected Factors of Socio-technical Systems Aggregated for the Management of Technical and Operational Aspects with the Greatest Impact on the Efficiency of Digital Business Models

No.	Socio-technical factor	Interpretation for the design of digital business models—digital platforms in the context of the different stages of research and application work
1.	Structural factors—related to the form of system components and their interrelationships.	**Conceptualization:** creating assumptions for the design of the digital platform based on the identified system components and the choice of technology. **Specification:** adoption of optimal solutions in the field of value creation logic and values obtained by means of specific technological solutions. **Operationalization:** indication of tasks performed by the system and the recipients of individual system functions. **Comparison:** selection of relevant databases and data collection modalities for the components of the digital platform. **Application:** identifying potential problems that may be encountered when designing the system.
2.	Behavioral factors—related to performance, operations and responses to underlying stimuli for the current "state of practice" of systems architecture and design and model-based systems engineering.	**Conceptualization:** analysis of potential user behavior is the proposed logic of value delivery and system operation mechanisms. **Specification:** defining customer service variables that need to be addressed. **Operationalization:** implementation of solutions to develop the issue of an attractive user service system. **Comparison:** testing hypotheses related to the proposed solutions against other solutions already used in their own applications or other suppliers. **Application**: implementation of actions revising the proposed ideas—verification and validation of solutions.
3.	Contextual factors—related to the circumstances in which the system exists.	**Conceptualization:** creating the concept of the business ecosystem and its cooperation with other ecosystems. **Specification:** defining variables responsible for the context in which the digital platform operates. **Operationalization:** proposing and implementing solutions to embed the platform in a specific business ecosystem. **Comparison:** verification of the proposed solutions compared to the requirements of the business ecosystem. **Application:** the formulation of specific problems to be addressed in the next stages of the design of the digital platform. Analysis of the proposed solutions.

No.	Socio-technical factor	Interpretation for the design of digital business models—digital platforms in the context of the different stages of research and application work
4.	Time factors—related to the dimensions/ properties of the systems over time.	**Conceptualization:** determination of the implementation time of the system and its life cycle. **Specification:** indication of precise dates and conditions for the implementation of the project, taking into account individual milestones. **Operationalization:** implementation of specific solutions in terms of the criterion of time management in the project. **Comparison:** verification of planned activities with real data from field work – design work in different locations. **Application**: defining factors responsible for time management and implementation of the various stages of the project.
5.	Perceptual factors—related to stakeholder preferences.	**Conceptualization:** identification and mapping of stakeholders interested in the functioning of the digital platform. **Specification:** decision-making setting of requirements for the design and the system to be used in the future. **Operationalization:** solutions related to managing relations with stakeholders. **Comparison:** verification of proposed solutions in relation to previously implemented projects and experiences from cooperation with stakeholders. **Application**: defining guidelines related to stakeholder relations in the context of a specific functionality of the digital platform.

Source: Own study based on Rhodes and Ross (2010) and Spendel (2005, p. 33).

studies of railway, energy and water and sewage sectors. The development of these characteristics was based on the assumptions of defining the areas of socio-technical systems such as the environment, goals/tasks, structures, leadership procedures, people and technologies. These areas proved that they meet the definitions of socio-technical systems. They clearly fit into the assumptions of this theory and provide the basis for application. It is also particularly important to note that they have incredible potential for embedding digital platforms in their ecosystems as part of business development and designing innovative business models that support these ecosystems in creating new functionalities. At the same time, they ensure the improvement of operational performance which is noticeable by customers as part of the restructuring of traditional companies towards digital and strategic renewal.

The expected cognitive objectives were achieved, including:

1. Identifying the factors responsible for shaping socio-technical systems.
2. Determining the relationship between the processes of designing digital business models based on digital platforms or technical infrastructure systems built from a network of links and the theoretical attributes of socio-technical systems.
3. Assessing the impact of social and ecological factors on the development of unique attributes of digital business models supported by popular social ideas.
4. Assessing the relationship between the conceptualization and operationalization of an innovative business idea through different theories.
5. Developing a new paradigm of strategic management that counters the unsuitable theories for the modern economy with concepts that do not fully work in the digital and global environment.

The same applies to the implementation of methodological objectives:

1. Identifying the factors responsible for ensuring coherence between socio-technical systems theory and the requirements for the design of digital business models and the digital transformation of technological companies.
2. Developing strategic recommendations for the process of scalability of solutions in the sphere of digitalization and taking into account different types and concepts of business models in the criterion of social impact thanks to technologies.
3. Developing a universal description of technological platforms grouping communities for the creation of economic value in the form of a business model (canvas).

4. Developing taxonomies of digital business models based on the assumptions of socio-technical systems.

As regards the key research questions, each of them was subjected to a detailed analysis which showed that:

1. There is a common denominator in the design of digital business models, where the genesis of this process is the use of the assumptions of the theory of socio-technical systems.
2. There is a synergy between the type of ideological approach adopted to define the social and managerial functions of the business model and its location in the digital economy in this network ecosystem.
3. The network paradigm is conducive to building complex systems built from technical infrastructure.
4. Technical compatibility and secure integration based on risk analysis and its need for approval in terms of matching two or more solutions of different manufacturers for the production of one product/system/subsystem/technology supporting the design and operation of digital platforms are crucial factors shaping the possibilities of entering into technology fusions that build a networked, digital business model embedded in specific legal, economic, social, political and ecological realities.
5. It was proven that companies use various methods in the field of management, aspects of creating innovations, including open innovations and technology fusions, as well as building network relations between systems and subsystems forming the business ecosystem, which is particularly important for the design of digital business models.

As indicated in the introduction, the relevant literature does not address the proposed topic to such a large extent. The value of the creative contribution is related to the presentation of a comprehensive approach to the types of digital business models, the unification of their description, and the inclusion of artificial intelligence, a network paradigm, business ecosystems and social impact, as well as safety and cybersecurity in the creation of innovative business solutions in this area.

7.8 STRATEGIC RECOMMENDATIONS FOR THE USE OF SOCIO-TECHNICAL SYSTEMS IN THE SHAPING AND DEVELOPMENT OF DIGITAL BUSINESS MODELS

As part of the research and scientific analyses, an analytical–evaluating study was conducted, the conclusions of which allowed for the creation of a set of axioms which prove the intensity of theoretical and practical knowledge in the field of a solved scientific problem related to the issue of socio-technical systems in the shaping and development of digital business models:

1. An important determinant in understanding the application of socio-technical systems is that both the technical and social worlds create the systemic world and are interdependent.
2. The space for creating digital business models based on digital platforms is to create social and technical systems and combine them in order to obtain the proper business environment. It can be called a business ecosystem based on the assumptions of integration of the social system with the technical system. Only the two systems combined create the conditions for creating new value for users and monetizing the business.
3. The five levels of the taxonomy of descriptors of socio-technical systems are equivalent in every dimension of the socio-technical system, namely in the social, technological and environmental areas. All factors, namely structural factors, behavioral factors, contextual factors, temporal factors and perceptual factors, are included in these three leading areas of system design. Each of these thematic areas should be integrated into a unified system based on the principle of complexity theory.
4. The concepts of the socio-technical system and business model are associated with the following concepts: digital platforms, ecosystems, innovation, competitive analysis, circular ecosystems, complex data, autonomous agents, technologies and evolutions, business and societies, technology development, agile manufacturing systems and metadata. These are very new concepts in the field of scientific and practical cognition. This therefore proves that the concept of socio-technical systems is increasingly being examined in the context of new trends in the digital economy and the design of digital business models.
5. Ecosystems allow companies to create new values that no company could achieve alone. A key role in this respect is played by social business policy, where communities and individuals are placed at the most important point of the designed businesses. Technological support serves people and the usefulness of the products and services they design. Social business policy is created by conscious business designers who pay special attention to the positive social impact.
6. The business ecosystem is based on the participation of large communities of people using IT applications designed on the principle of the technology fusion of key players in digital business.

7. The scalability of social enterprises is an extremely interesting concept as it involves a situation where individual initiatives affect a very wide audience. Then, a relationship referring to the concept of a multiplier of the scale of operation of social enterprises occurs. When a given initiative significantly affects the social environment and the social environment needs this initiative for its development, the scalability of a social enterprise can be referred to. When the social impact significantly exceeds the expenditure transferred to the implementation of a specific social program, the scalability effect is achieved.
8. The relationship between the resources involved in the production or provision of services and the ecological effects should be significantly positive. In other words, the added value will not only be the effect of a linear value chain, but of a circularly oriented process value loop. In this respect, the single process of creating added value should be multiplied by further possibilities of reduction, reuse, recycling and other mechanisms of this concept. If the ecological effect of a single company is combined with other companies operating in the same sector and in related sectors, circular business models will jointly strive to create ecological value influencing positive climate change for the wider public, and thus, as it were, society as a whole.
9. Green digital business models are characterized by two reasons why they can be called such. Firstly, digital technology itself is based on optimizing the use of natural resources. Secondly, these business models are focused on creating ideas supporting the implementation of climate policy that is positive for nature. Green digitalization therefore has built-in attributes, both in technology and in popular ideas attractive to environmentally oriented consumers. These two areas make it possible to create such defined green digital business models.
10. Technical scalability depends on the elements of the technical system together with feedback loops in the context of given tasks for specific functions of that system. Economic scalability is conditioned in terms of the relationship between a revenue growth rate and the cost structure depending on this process. Social scalability, in turn, determines the dependence of the number of users of the digital business model who influence the social system within the limits of its impact.
11. Social impact is crucial in the context of designing scalable business models, including those that are ultimately to affect climate change. In addition to the economic factor, a social factor

also occurs, where people even demand clean air, clean water and an aesthetic and clean architectural or natural environment.

12. The concept of the green scalability of circular business models shifts the focus from the scalability of a single company to scalability which will benefit not only that company, but also a wider audience – the wider society. Scalability defined as such increases the positive social and ecological impact of the business.
13. The core determinant of the construction of socio-technical systems is connecting the human community with technological factors. The modern challenges of business digitalization mean that there are a number of technological solutions around people's lives that fit into the context of people's functioning in the professional and life ecosystem.
14. Theoretical potential in the assumption of a combination of social factors taking into account the role of humanism in the design of organizations and the potential embedded in innovative technologies favors the possibility of describing digital business models.
15. Nowadays, in the era of digital transformation, innovations built from socio-technical systems have been appreciated and their achievements are useful in designing new business solutions. The complexity of technology-based business systems is a crucial aspect of examining socio-technical systems. Complex systems, such as socio-technical systems, require detailed research, analysis and assessment of their applicability in the business environment. Man versus machine, from the cybernetic perspective, is a leading topic on the border of management, engineering and sociological research.
16. The social ecosystem of business is one in which not only the interests of the individual are assessed, but in particular the interest of the general public. The digital platform is intended not only to serve the needs of these individuals, but to provide added value to a large group of users and stakeholders.
17. The key issue of understanding the concept of the business ecosystem is the virtual environment based on digitalization. In this sense, this phenomenon creates the concept of ecosystems. Then, it is essential to create a focus on success supported by digital solutions in areas such as customer focus, integration of multiple value chain ecosystems, Big Data analysis, real-time technologies and other technical innovations.

18. The network creates business models based on the network potential. The better the effects of networking, the greater a chance to create innovative business models. The network thus creates a business ecosystem.
19. A lack of implementation of pro-social orientation in many cases is a barrier to the development of modernly managed companies. In turn, the lack of progress resulting in technological debt pertaining to a failure to implement modern solutions reduces the level of competitiveness of companies and does not bring them to the level of business platformization, i.e., the transfer of businesses into the space of complex systems cooperating with other complex systems based on technology fusions.
20. The network of relationships is the foundation of the effectiveness and efficiency of network business models. Network business models are solutions in which the condition for creating the expected function is the coexistence of at least two other business models with the consistency of technological, organizational and business aspects. The network and its attributes determine the competitiveness of the network business model. In turn, the technological aspect used in the environment of complex systems is focused on pursuing the purpose function in business and operational terms based on processes and projects.
21. The business model based on the digital platform can be powered by artificial intelligence algorithms that support its traditional functions implemented by means of the technologies used. Learning of the business system – the degree of value delivery matching – is improved through dynamic feedback loops based on the assumptions of cybernetics theory.
22. The theory of cybernetics should be taken into account in the context of the determinants of designing digital business models, while also taking into account the assumptions of socio-technical systems. The examination of cyber solutions in the field of network theory is becoming particularly important. In this regard, building complex systems connected by interfaces and controlled centrally, and in some cases decentralized, should be supported by knowledge and experience in the field of cybernetics.
23. Complex systems meet the definition of the System of Systems concept, which is based on the interdependent creation of a multidimensional and interdisciplinary system of network connections, where elements are mutually dependent and dynamically affects each other. Adaptive complex systems can

both influence their environment and change their structure without external input.

24. Socio-technical systems that underlie the construction of business platforms are based on a modular approach, where their complexity, consistency and ingenuity determine the success or failure of the proposed solution. Stimulating disruptive ideas in sectors are geared towards specialization, on the basis of which companies build competitive advantages and in many cases create new blue ocean business space.
25. The design engineering of digital platforms should be combined with softer elements, which include aspects of socio-technical systems. A balanced focus on the human perspective and the technology perspective helps design solutions that, on the one hand, are attractive ideas acceptable to large human communities, but, on the other hand, are easy to use, widely available and socially acceptable.

7.9 CONCLUSIONS

The author's original solutions presented in the chapter demonstrate that social and technical systems provide a platform for building modern business solutions based on the digital economy and, in particular, digital platforms. The four business ecosystems demonstrated, namely the digital platform ecosystem, the rail transport ecosystem, the water and sewage business ecosystem and the energy sector ecosystem, demonstrate the need to build digital business models supporting digital transformation and product digitalization processes in these areas. It was proven that digital platforms are designed in specific sectors of the economy and serve not only the core areas of citizens' lives, but in particular, are used as specialist tools for process improvement in specific sectors of the economies. It should be expected that the most revolutionary solutions that will change the conditions of the game in these areas of business will be created in these areas. In this chapter, the author's original canvas of the description of digital business models was developed, taking into account complexity theory, the assumptions of socio-technical systems and business digitalization. An original methodology for designing digital business models, which should support designers in their daily work and pay attention to those areas that they could omit to create excellent business solutions, was also developed. The discussion on management factors in the area of digital transformation should give rise to the practical application of the monograph content.

REFERENCES

Appelbaum Steven, H. 1997. Socio-technical systems theory: an intervention strategy for organizational development. Management Decision. 35–36: 452–463.

Blaschke, M., Cigaina, M., Riss, U.V. and Shoshan, I. 2017. Designing business models for the digital economy. pp. 121–136. *In*: G. Oswald and M. Kleinemeier (eds). Shaping the Digital Enterprise, Switzerland: Springer International Publishing. DOI: 10.1007/978-3-319-40967-2_6.

EN 50126-1 Railway Applications—The specification and demonstration of Reliability, Availability, Mantainability and Safety (RAMS)—Part 1. 2021. Generic RAMS Process and T. Myklebust, T. Stalhane, Function Safety and Proof of Compliance, Springer Nature Switzerland. 54.

European Committee for Electrotechnical Standardization—European Committee for Electrotechnical Standardization is responsible for European standardization in the area of electrical engineering.

Hoehne, O. 2016. Rail systems viewed from a system of systems perspective. A publication of the International Council on Systems Engineering. 19(3): 36.

Kreft, J. 2021. Władza platform. Za fasadą Google, Facebooka i Spotify, Universitas, Kraków. p. 25.

Osterwalder, A. and Pigneur, Y. 2010. Business model generation: a handbook for visionaries. Game Changers and Challengers. Wiley, Hoboken, NJ.

Rhodes, D.H., Ross, A.M. 2010. Shaping socio-technical system innovation strategies using a five aspects taxonomy. Massachusetts Institute of Technology.

Spendel, Z. 2005. Metodologia badań psychologicznych jako forma świadomości historycznej [Methodology of psychological research as a form of historical awareness]. Wydawnictwo Uniwersytetu Śląskiego, Katowice. 33.

Conclusion

The content presented in this monograph proves the complexity of the processes of designing and developing digital business models operationalized in the form of digital platforms. In theoretical terms, theories such as complexity theory, socio-technical systems, production and service systems and configuration management in the process of designing business models were used. Digital business is growing extremely dynamically. The number of startups based on digital platforms launched on the market is very large. However, not all of them will be successful. The process of designing digital business models must be focused on success, and this is conditioned by the attractiveness of the digital business model, the social acceptance of potential recipients and the impact of these platforms on the quality of life of citizens. The core element is also whether the proposed solution crosses the border of local markets and becomes a global solution. The digital business model must generate tangible benefits for users and this should be embedded in its logic in the sphere of value delivery, but also in the issue of gathering large communities around itself, aggregating Big Data sets and accessibility. Digital platforms enable the sharing of IT infrastructure resources and fusions based on the exchange of technologies supporting the designed solutions. Platforms connect markets, users, manufacturers, suppliers and various stakeholder groups. In a way, business as a whole moves to virtual solutions—digital ecosystems. In this way, innovative solutions are created that satisfy new customer needs or enhance current ones in a more comprehensive way. At the same time, ground-breaking innovations are being created to change the conditions in the sector—the game changer phenomenon. The diffusion of digital technologies creates new space in digital business. Socio-technical systems that derive from the assumptions of the theory of complex systems and basic principles of business management imposed on the layer of production

DOI: 10.1201/9781003456421-8

of new innovations based on digital platforms draw attention to the joint implementation of community creation and the use of modern technologies for this purpose. Technologies are therefore a condition for connecting people, creating interactions and building a communication bridge between them. Such a possibility did not exist in the traditional approach to management. Paying attention to socio-technical systems helps us understand the principles related to the design of digital business models and, in particular, digital platforms better. Focusing on community building and selecting the appropriate technology for the process makes it easier for designers to create innovative and market-attractive solutions that can succeed. This monograph tries to look at digital platforms in a multidimensional way from many perspectives. Such an interdisciplinary approach based on explaining theoretical issues, the conceptualization of proposed solutions and the operationalization thereof by indicating various solutions in chapter seven that can support managers in designing digital business models, but also in their management, is the aim of this monograph. An attempt was made to explain complex processes, concepts, definitions and taxonomies of concepts in order to create the best value for the reader so that this monograph would become a guide to understanding the modern digital economy. The creation of digital business models is not easy. Many projects fail. The monograph explains how to approach this type of project and what the relationships between numerous variables that affect success or failure and define various threats, limitations and directions of development of this type of business are. The author hopes that the monograph will be warmly received by readers and will add value to current knowledge in this complex matter.

Index